Insights
into the
Scriptures

The Book of Mormon

The Jaredites

Figure 1: The Tower of Babel

By
Daris Howard

The Book of Mormon – The Jaredites
Insights Into the Scriptures
By
Daris W. Howard

Copyright © 2023
by
Daris W. Howard

All rights reserved. No part of this book may be reproduced or transmitted in any form or by any means, electronic or mechanical, including photocopying, recording, or by any information storage and retrieval system without the written consent of the publisher.

ISBN-10: 1-62986-028-X
ISBN-13: 978-1-62986-028-2

www.publishinginspiration.com

Publishing Date: May 4, 2023

Publishing Inspiration LLC

Table of Contents

Thanks and Recognition

I would like to offer my gratitude to all who have willingly reviewed and given me feedback on this book. First and foremost is my wife, Donna, who has constantly helped me refine this book. Others (ordered alphabetically by last name) include:

Scott Howard
Gavin Howard
Melissa Moses Lyman
Dr. John L. Walker, PhD

Dedication

I dedicate this book to my daughter, Jenna. We used to have wonderful discussions in the evening as she studied scriptures, and some of the insights I gained came from questions she asked that I was able to ponder.
Daris Howard

1
What is the Book of Ether?
+

This book is intended for those who have an understanding of The Book of Mormon, and in particular, are familiar with the book of Ether. For that reason, this chapter was not originally planned for this book. However, on the suggestion of reviewers that there may be those interested in this book for the purpose of considering ancient cultures, this chapter has been added.

So, what is the book of Ether? The book of Ether is a section of The Book of Mormon, much like, for example, Joshua is a book in the Bible. But how it fits into The Book of Mormon is slightly different. Before we can speak of the book of Ether, we must understand how The Book of Mormon came into existence, its purpose, and a general consideration of its content.

Just as the Bible was written by ancient prophets in the Middle East, each writing their own part, The Book of Mormon was written by prophets on the American continent. As a young boy in his fourteenth year of age, Joseph Smith sought to know what church was true since every Christian church in his region interpreted the Bible differently. To Joseph's inquiry, he received a heavenly vision.

From that point, much as Samuel in the Old Testament, Joseph was tutored by the Lord. Just as with Samuel, this youthful training helped Joseph avoid being overly influenced by the culture surrounding him and prepared him to be an instrument in the hands of the Lord.

After a few years, Joseph received a heavenly visitation and was told about the gold plates on which The Book of Mormon was written. Though he saw them and where they were deposited, he was not allowed to take them at that time. He received four more years of training, maturing in the ways of the Lord, before he received them. By the use of a Urim and Thumim, prepared by the Lord to do translations just as prophets did anciently (see Exodus 28), Joseph was able to do the translation.

At the time that Joseph did this work, most Christian churches claimed revelation had ceased. But as Paul told us in his letter to the Hebrews in the Bible:

Hebrews 13:8 Jesus Christ the same yesterday, and to day, and for ever.

1

As the Lord revealed His word to prophets in the past, He would do so in modern times as well, as long as there were those worthy and willing to receive it. We only have to look at the world around us to know how much it is needed.

But as can be imagined, many were skeptical of Joseph's claims of the divine way in which The Book of Mormon came. The Lord, therefore, prepared other witnesses of the work. Three men received a revelation from the same angelic messenger who first told Joseph about the golden plates. Many others saw and felt the plates, though they received no angelic visitation.

These witnesses were told to share their experiences with the world. They have done so in the opening pages of The Book of Mormon. Many of these men fell away from the church Jesus Christ directed Joseph to organize, but still never varied from their testimony of their experience related to their witness of The Book of Mormon.

That brings us to the question of the purpose of The Book of Mormon.

As Joseph found, there were many interpretations of the Bible. It was much like a board with a single nail in it. The board can be rotated around the nail to infinitely many positions. The Book of Mormon became like a second nail in the board, locking in a straight path between points of doctrine. This fulfills the prophecy in the Bible given by Ezekial in chapter 37, that there would come together the scroll or "stick" of Judah (the Bible) and the stick of Joseph (The Book of Mormon).

16 Moreover, thou son of man, take thee one stick, and write upon it, For Judah, and for the children of Israel his companions: then take another stick, and write upon it, For Joseph, the stick of Ephraim, and for all the house of Israel his companions:
17 And join them one to another into one stick; and they shall become one in thine hand.
18 And when the children of thy people shall speak unto thee, saying, Wilt thou not shew us what thou meanest by these?
19 Say unto them, Thus saith the Lord God; Behold, I will take the stick of Joseph, which is in the hand of Ephraim, and the tribes of Israel his fellows, and will put them with him, even with the stick of Judah, and make them one stick, and they shall be one in mine hand.

As these two books, the Bible and The Book of Mormon, became one, they made the understanding of the Lord's words clearer and more well-defined. Also, as doubt about God in our world increases, The Book of Mormon gives a second witness to the divinity of Jesus Christ and his role in the salvation of God's children here on earth.

This brings us to the final question: What is the content of The Book of Mormon, and how does the book of Ether fit into it?

The Book of Mormon mainly follows two migrations of people to the Americas, which they called "The Promised Land" because of the Lord's promises regarding it. The core part of The Book of Mormon contains the story and work of the Lord with a group that came out of Jerusalem at the time Zedekiah was the king (around 600 BC).

There were many prophets in, around, or who came out of Jerusalem at this time, including Isaiah, Ezekiel, Hosea, Amos, Micah, Zechariah, and Daniel. Probably the most famous prophet of the time was Jeremiah. Some of these prophesied and died in Judah. Others were taken and prophesied in exile in Babylon. Many of these prophets preached about the wickedness of the people and that if they did not repent and submit to the Babylonians who had conquered them, they would be destroyed and taken captive to Babylon when the Babylonians returned. (The Babylonians had already subjugated them once.)

One of these prophets The Book of Mormon tells us about was a man named Lehi. He and his family were commanded by the Lord to leave Jerusalem and eventually build a ship and migrate to the land of promise. Once in the Promised Land, those in this party splintered into two main groups, the Nephites and the Lamanites, named after their leaders. These two nations were often at war with each other, much as the Israelites and their neighbors in the Bible were. The Book of Mormon bears the writings and prophecies of these American prophets, containing in its pages doctrines to clarify and substantiate the Bible.

The culmination of these writings comes at the time of the birth and death of Jesus Christ, with the pinnacle being His visitation among the people after his resurrection in Jerusalem. The record of this people ended with the destruction of the Nephites around 400 AD when the majority of them turned wicked. The record of the groups from this migration is about 500 pages of the approximately 530 pages of The Book of Mormon.

Most of these 530 pages were not compilations of multiple prophets' records combined into one like the Bible was. Instead, most of the book is the abridgment of these prophets' writings done by Mormon, one of the final Nephite prophets. It is because of his work that The Book of Mormon bears his name.

The other thirty pages of The Book of Mormon is a record of a separate people that migrated to the American continents much earlier. They came from The Tower of Babel, which would put their departure to the Americas around 2000 BC (the approximate time of The Tower of Babel). In modern times, they are often called Jaredites, a name taken from one of their first leaders. This name does not appear in The Book of Mormon, but is used by most readers of the book to distinguish them from the Nephites and Lamanites.

Their record came to be part of The Book of Mormon when the Lord helped the Nephites find twenty-four gold plates on which the Jaredite record was written. These gold plates were written by a prophet named Ether, from which this section of The Book of Mormon derives its name. The twenty-four gold plates were not directly part of The Book of Mormon, but were abridged into it by Moroni, Mormon's son.

This record of Ether covers a time period earlier than most of the rest of The Book of Mormon, and hundreds of years longer. Because it is taken from a people much more ancient than that of the Nephite and Lamanite nations, the culture is far different. The book of Ether is the abridgement of the prophetic writings about this people.

This book is by no means meant to replace a study of the book of Ether. It is intended to give significance, understanding, and insight into a culture far removed from most cultures of modern times. If you, the reader, have not read the book of Ether, I suggest you do so. I also suggest you read The Book of Mormon in its entirety, too.

Many who start The Book of Mormon never finish it. Most of those who stop reading fail to get through sections where the prophet Nephi quotes from the book of Isaiah. Isaiah can indeed be a tough read for those of us who are not familiar with the prose and culture of that Biblical time period. There is even a story about a man who kept a miniature copy of The Book of Mormon in his pocket when he went to war. One day, he was shot, and The Book of Mormon saved his life. The bullet stopped in the Isaiah sections. The man quipped that even a bullet can't get through Isaiah.

Humor aside, if those sections bog you down, save them for later, and read beyond them. There is much to be gained from this study.

And with that introduction, let's proceed to insights into the magnificent culture of the Jaredites.

2
Why Write This Book?
✦

I considered making this chapter a forward or prologue instead, but too many people seem to skip those. I feel it is essential to understand the reason for this book and how I designed it, so I chose to make it a chapter. Therefore, I present the reason for the book here.

I had just finished a presentation on the Jaredites at an Education Week class at BYU – Idaho when people came up to speak to me. The last person was a lady.

"Brother Howard, have you ever considered writing a book about the insights into the Book of Mormon that you shared with us?" she asked.

"No," I replied. "Some research and ideas are original to me, but probably at least half of everything I have shared with you can be found in the works of Hugh Nibley, Thor Heyerdahl, the Neil Maxwell Institute at BYU, and other places. I may have discovered some things or looked at them differently than others, but I'm not sure I feel it will fill a book."

"It's not just what you shared," she said. "It's how you shared it. I've tried reading Nibley and some others, but I get too bogged down. You shared it in ways, and with stories and methods, I can relate to. You also kept it short and to the point. I really think you should consider writing the material into a book."

After my conversation with her, I wondered if there was something I could share that might be useful to others. I considered that many scholars, all probably smarter than I am, have written on many of these topics. As I pondered this question, I also considered an event that had happened to me shortly before an Education Week presentation.

A man and I started into what I thought would be a pleasant academic discussion about the Book of Mormon and its geography. The man I was visiting with suggested that he would share his ideas, after which I would share mine. After he had gone through his reasons for where he thought The Book of Mormon lands were, and it was my turn, he asked if I agreed with all he said. When I told him I had views that differed from his, he said, "I can see that this will just create contention." Then he walked away before I could share my viewpoint.

I must admit that I was somewhat annoyed by this—almost angry. I had listened to him and wanted to share my ideas as well. I was also taken aback because I had thought we would have an academic discussion, but he felt only his side was worth telling. He seemed to feel mine was useless and would cause contention.

As I considered this over the following week, I felt strongly that the spirit was rebuking me, telling me that there was something I should learn from this experience. I must admit that my intense annoyance with what had happened made me less than happy about having instruction from the spirit, especially in the form of a rebuke. But when I finally humbled myself enough to ask what He would teach me, I felt Him tell me some interesting things.

The spirit told me it was okay to consider scriptures academically, but I needed to understand that it was not the most important part of any religious discussion. I already knew this from my experiences in the mission field. No one I shared the gospel with ever gained a testimony from academic instruction alone. Supposed academic discussions usually turned into nothing but a contest of who knew the scriptures best and could apply them to his side of the argument.

I felt that the spirit was saying something even more profound. It doesn't matter who is right and who is wrong, or if anyone is. But what really matters is the learning process and being open to the spirit to give each of us a chance to enlighten our understanding and strengthen our testimony. That is far more important than any academic knowledge that can be gained. Even in academic studies of the scriptures, the crucial part is to be open to lessons that the Lord would have us learn.

Nephi said:

O that cunning plan of the evil one! O the vainness, and the frailties, and the foolishness of men! When they are learned they think they are wise, and they hearken not unto the counsel of God, for they set it aside, supposing they know of themselves, wherefore, their wisdom is foolishness and it profiteth them not. And they shall perish.

But to be learned is good if they hearken unto the counsels of God. (2 Nephi 9: 28-29)

President Hinckley said:

"What did Paul mean by the words, a sound mind? I think he meant the basic logic of the gospel. To me the gospel is not a great mass of theological jargon. It is a simple and beautiful and logical thing, with one quiet truth following another in orderly sequence. I do not fret over the mysteries. I do not worry whether the heavenly gates swing or slide. I am only concerned that they open. I am not worried that the Prophet Joseph Smith gave a number of versions of the first vision anymore than I am worried that there are four different writers of the gospels in the New Testament, each with his own perceptions, each telling the events to meet his own purpose for writing at the time.

I am more concerned with the fact that God has revealed in this dispensation a great and marvelous and beautiful plan that motivates men and women to love their Creator and their Redeemer, to appreciate and serve one another, to walk in faith on the road that leads to immortality and eternal life." [1]

As I felt the direction of the spirit telling me that I might be too caught up in academic concerns, such as some people seem to be with the exact location of Zarahemla, I realized that there were more important, spiritual things that I needed to learn. Even in an academic approach to the scriptures, the journey the spirit could take me on was the essential part.

I also felt the spirit tell me that God had chosen not to reveal things we might desire to know to test us to determine if we will trust him for the things he has left unanswered.

I have academic friends who have felt they could prove the gospel through study and research. It has been my experience that this is impossible. God has left holes that he has chosen not to fill. He has done this so we have to live by faith. Faith is one of the fundamental principles upon which the work of God is built and one upon which we must have a firm foundation to return to Him. In a strictly academic approach, eventually a person will hit those holes. If that person's foundation is not built on the rock of faith and trust in God for what can't be learned or determined academically, then, often, the foundation will crumble under the weight of academic doubt.

I realized that academic pursuit is not wrong, but its usefulness is for gaining a deeper understanding and the faith God would have us gain. And where the spirit might lead one person academically, it might lead another person in a different, seemingly opposite, direction. But what the spirit will teach in the process is the important thing.

A second idea I felt the spirit say to me dealt with the concept of academic understanding. I felt the spirit tell me that the exact location of events in The Book of Mormon is not as essential as the spiritual journey. I wondered why, then, it wasn't revealed where those places were, and the questions could be put to rest instead of leaving us searching. As I pondered this, I felt the spirit tell me God chose not to disclose details that would give us exact locations on purpose. I felt him say, "God does not need a Mecca for his people to worship. They need to focus their worship on God and Jesus Christ."

Is it possible that if we knew the exact location of Zarahemla or the place of Christ's appearance to the Nephites that those places would draw us there, and we would come to possibly worship the place more than we worship the true and living God?

Don't get me wrong; I served my mission in New York and love visiting places like The Sacred Grove. I love to feel the spirit there as I consider the incredible event that occurred on that glorious day in the Spring of 1820 when God the Father and Jesus Christ appeared to Joseph Smith. I have taught prospective converts there and prayed there. But if a person considers the physical place as the focus of worship, he has lost the overall picture of the reasons for the beautiful plan of salvation and its goal of bringing us back to God's presence.

However, if in academic study, the Holy Ghost can teach us spiritual lessons that will make us more like our Savior, then the study is worthwhile. If we can gain an understanding of people, ancient and modern, and how their lives have been brought closer to God by the things they have experienced, then perhaps we can incorporate those insights into our own lives. This type of academic study would be beneficial to us spiritually.

For that reason, as I have worked on this book and other religious books, I have chosen to emphasize the spiritual journey and learning the spirit has guided me on as much as any academic ideas I might put forth. Some of these included rebukes I have felt from the spirit as I was shown my weaknesses. Though I do not expound to the world my shortcomings, I have not tried to hide the rebukes I have felt.

I admire Joseph Smith's willingness to share sections in the Doctrine and Covenants when God rebuked him. If I had been in his shoes, I might have said, "I think this is one section that doesn't need to be printed." But he was humble enough to accept the Lord's rebuke and allow us to learn from it.

I doubt that every reader will agree with all the ideas I put forth from research in archaeology, science, and other fields. But I hope my spiritual growth and journey can benefit truth seekers as they approach their own scriptural exploration. In fact, in some ways, I hope that everyone finds a few things they might disagree with so each person can follow their own spiritual adventure and seek for their own understanding to learn what the spirit will teach them.

With that said, I hope that my work and research might open some ideas of understanding for each reader. I hope it will be a place for readers to begin or continue their own personal investigation. But I also do not want anyone to take my work as absolute, proven truth. There is only one source from which truth comes that is infallible, and that source is the divine. God only designates a few chosen men to speak with that divine authority for the world, and I am not one of them.

Also, this book is meant to supplement the reading of the scriptures, not replace it. The scriptures, as divine writings, carry a spirit to help people answer their questions. No academic writing can supplant that.

I also wanted to keep the book small, succinct, and in as readable of a form as possible. I am a syndicated writer, writing short stories for newspapers and magazines across the U.S. and parts of the world. I finished a doctorate not long ago, and my wife sometimes comments that my writing can take on a stiff, academic style when I am not careful. I did not want that here any more than necessary. I have included stories of my journey to hopefully make the reading more understandable and fun for the reader.

I do want to provide references for the reader to look further if desired. Still, I am not adhering to the standards of my dissertation that only scholarly works that are extensively expert-reviewed should be used. This would exclude newspapers or magazines such as National Geographic, which often have much to consider. I know this may not satisfy the strict academic, but I am not trying to make a book that would pass the dissertation muster, but one that will be enjoyable to read and raise ideas the reader can consider.

On the other hand, I am also trying to write a book that pulls in information from experts and has depth to it. Therefore, it needs to have at least a taste of research orientation. In doing both of these things, I may not please either the research-oriented person or the casual reader, but I hope it will be of some value to raise ideas that have yet to be considered.

I also wanted to keep the book's topic narrow and as clear and simple as I can make it; a book that can be read in a couple of Sunday afternoons. My Education Week presentations start with a list of questions to answer, and I have chosen to do the same with this book, making these the chapters. In a somewhat unusual method for a book, I made these questions the table of contents and put them in outline form, much like a research paper. This should help readers look up specific ideas more efficiently.

Though my doctoral work used the APA style of reference, reviewers suggested that the style of this book lends itself to a much less obtrusive reference for reading style. For that reason, I have chosen to use the IEEE reference form with small numbers in brackets, like [#], indicating items in the bibliography.

Where possible, I have done one last thing. In the electronic version I have put links to websites near the related reading material if links are available. These links are indicated with [WL] for web link. This is more of a modern technology idea than an academic one. The only reason I enjoy reading scriptures or study material on my phone over a physical book is that I can quickly click a related link and read additional material. One challenge with reading from a physical book is that if I go to the link, when I return I can't remember where I was. (Maybe it is because of my age.) Of course, those reference links won't be clickable for a person reading a physical book copy, so they will be removed.

Finally, I hope you, the reader, are ready to embark with me on a world of discovery and spiritual growth, questioning, wondering, and growing as you work toward the values and ideas the spirit would have you obtain for yourself as much or more than any academic ideas we can share together. And I hope it will be an enjoyable expedition.

Daris Howard

3
Why Were Scholars Originally Critical of the Book of Ether?

✦

When I was a boy, church meetings were laid out differently than they are today. There was Priesthood and Relief Society in the morning. After that, there was Sunday School. We would go home around lunchtime and return for sacrament meeting in the evening. While my parents and brothers were in Priesthood and Relief Society, my sisters would read to us from scripture readers for the Book of Mormon. The main book I loved was the book of the Ether—the Jaredites. I was not sure why other than to say that their story was different enough that it fascinated me. As a boy, I couldn't even say why it was different—it just was.

As I became older and was able to read the Book of Mormon for myself, my interest in the Book of Ether never decreased. In fact, if anything, it increased. There were so many questions about it that I wanted to answer.

Later, I served in the New York, Rochester mission. I met people there who were descendants of those who knew the Smith family. Some of them said there was admiration expressed in the journals of their ancestors for the Smiths. Others had strange, unbelievable ideas about them.

But one person shared something that made me think again about the Book of Ether. He said that the book of Ether was strange, and no one could believe it. He said his ancestors, who were scholars, felt the same way. His statement made me decide to look deeper, to determine what about the Jaredites was different, and to somehow put it into words.

On my mission, I became deeply familiar with the Jaredites by studying the Book of Mormon. But when I came home, I began to consider the strangeness the man had expressed about this book. Some ideas he shared gave me a starting point for my search. I decided to compose them in terms of questions and attempt to answer them. I realized the questions were mainly related to culture, and primarily due to what we might call unusual behavior in war. I have listed them here and will use them as a guide. There are other questions that people have raised in my Education Week classes at BYU - Idaho, and I will put them later in the book. These original ideas are only briefly introduced in this chapter, along with why they may seem unusual to us. Cultural reasons for them will be discussed in later chapters. The initial questions are:

1. When the Jaredites went to war, when was the war over?
2. Did the Jaredites prefer to kill or capture the leader of the opposing side?
3. Does it take an entire army to win the war?
4. What did the Jaredites do with captured leaders?
5. What did the ordinary people do once they got new leaders?

For each of these five questions, in this chapter I will briefly introduce the idea, share some comparisons between the Jaredites and our modern western cultures, and then talk briefly about scholar skepticism.

When the Jaredites went to war, when was the war over?

When the Jaredites went to war, the war continued until the leader of one of the sides was killed or captured. This was true in all times of battle, not just in the final destructive battle where only Coriantumr was left. Interestingly, this strategy can be seen in the game of chess.

In chess, everything is done to protect the king, with all other players being significantly less important, no matter how powerful they are. Every other player can be sacrificed to protect the king. We'll discuss chess more later. But an example of this can be found in Ether 8:

Ether 8:6 And it came to pass that when they had slain the army of Jared they were about to slay him also; and he pled with them that they would not slay him, and he would give up the kingdom unto his father. And it came to pass that they did grant unto him his life.

The whole army of Jared was killed fighting to preserve the king. But only when the king was dead or captured would the war be over. One side can lose everyone or almost everyone and still win if the opposing king is killed or captured.

In Ether 15, we read about Coriantumr:

Ether 15:4 And it came to pass that he wrote an epistle unto Shiz, desiring him that he would spare the people, and he would give up the kingdom for the sake of the lives of the people.
Ether 15: 5 And it came to pass that when Shiz had received his epistle he wrote an epistle unto Coriantumr, that if he would give himself up, that he might slay him with his own sword, that he would spare the lives of the people.

This was the only unacceptable thing Shiz could ask. Those who followed Coriantumr could not allow him to fall before they had given their lives to save him. And thus, the war continued and would until one of the two leaders was killed or captured. This is true whether it took an entire army to kill or capture the king, or whether the king was killed or captured by a handful of men, or even just one.

Comparing this to our western culture, can you imagine a man who wants to take over our government sending in a commando squad to kill or capture our president? Maybe the man was the leader of Al-Qaeda, ISIS, or some other organization with which we have a conflict. Or perhaps it is just some man who has become influential through wealth or a popular movement. Let's say the commando squad this man sends in somehow succeeded in their assignment. Would that signal the end of the conflict with them?

The thought is quite laughable to us. If anything, it would renew our efforts to get the perpetrators and punish them. This would be true of every Western Civilization, whether it was Europe, the Americas, Australia, or any other continent. If someone came into that country and killed or captured the president or prime minister, it would not signal the end of the conflict but would create a redoubling of effort to bring those responsible to justice.

It is easy to see why scholars in the days of Joseph Smith mocked this idea. Though the presidents, prime ministers, or other rulers in the respective Western Civilization countries are important, they are not the ultimate defining person who determines a nation's rise, fall, and stability. The succession of the country's government would follow some pre-designed rule.

For example, if something happened to the president in the United States, the vice president would then become president. If something also happened to the vice president, the next in line would be the Speaker of the House, then the President Pro Tempore of the Senate, and so on, as outlined in succession order.

Did the Jaredites prefer to kill or capture the opposing leader?

This is an interesting question. Sometimes the fight would go on until the leader was killed. This is especially seen in the final battle, which continued until everyone was dead except for the leaders of the two sides, Shiz and Coriantumr. And then the war was finally finished as Coriantumr took Shiz's life. But many of the wars did not have such extreme outcomes. Some ended with little or no bloodshed.

In the first battle, King Kib's son, Corihor, raised an army and fought against him (Ether 7). Corihor captures Kib but does not kill him. Instead, Kib remains in captivity until he is old. At that point, another son, Shule, born in captivity, raises an army and defeats his brother, Corihor, to return the kingdom to his father.

Note that Corihor is not killed, and when Shule is made the new king, Corihor repents and is given power in the kingdom. This makes two situations within the rule of the first generation of kings where the defeated opposition leader is left alive and subject to the king. This idea of not killing the king but putting him in captivity is mentioned eleven times in the book of Ether. *(See Appendix A on the table of king succession.)* In comparison, only ten times was the leader killed by the opposing leader, and three of these were by Coriantumr in the final war of the Jaredites.

Besides the ones explicitly mentioned, there are others where it is unknown if the king is killed or captured. However, the number of known times the opposing king is spared is on par with how many times the opposing king is killed. This means the desire to capture the opposing king and not kill him is significant and might even be preferred over killing him.

Does it take a whole army to win a war?

One important concept is that it does not take an entire army to win the war. Often there is what we might call a coup, which does not require a full-scale military campaign. An example of this can be seen in the changeovers of power from Omer to Jared to Akish.

This type of transition of power by a coup is familiar in our day, but there is something that makes it quite different. In our day, most transitions of power by coups are done by the military or have the backing of the military. This may not always be the case, but a person would be hard-pressed to take over a country if they were not somehow aligned with the military.

If a person were to attempt a coup and the military did not support them, it would be nothing for the military leaders to turn against them. Coup leaders would have little ability to stand against the military's might.

Now, contrast that to some of what could be considered coups in the book of Ether. Perhaps they could be military leaders, and it is not stated, but some situations give reason to believe that is not the case. Jared, in chapters eight and nine of Ether, is a good example of this. Jared, the son of King Omer, raised an army and became a military commander. As a military commander, he fought against his father and beat him, placing him in captivity. Later, Jared's brothers, born in captivity, raised an army and beat Jared. It says they destroyed his entire army.

Now, here is where the story breaks from the modern day. Jared now has no army but still desires the kingdom. This is when his daughter suggests she will dance for Akish and get Akish to kill Omer, and Jared will become king.

Jared has no army to back him and has just been defeated by the army raised by his brothers. He is, if you will, an enemy to his brothers, the army leaders. He is also guilty of insurrection and treason. The Lord warns Omer to flee, and Akish is able to put Jared in as king. (Akish, of course, later kills Jared and takes the kingdom for himself.)

The question arises as to whether Akish had an army behind him. If so, how did the army change so quickly from backing Omer and his sons, who defeated Jared, to supporting Akish, or was it a different army, and how did he organize it? Also, why were the people so willing to follow someone who had been guilty of treason?

Another example is someone like Heth in Ether 9:26-27. Heth's father, Com, was the king. It says Heth slew his father with his own sword and became king. Was this done with an army or by him killing his father and declaring that he was the king? It appears to be the latter.

So, consider why this would seem so strange to scholars at the time of Joseph Smith. What if someone were to go in and kill the leader of our country and declare himself to be the king? We wouldn't put up with that for a second and would demand the military leaders go after him. This brings us to another question. What did the ordinary people do when a new man declared himself to be king? We will visit this question in a moment.

What did the Jaredites do with captured leaders?

When the opposing side's leader was captured, he was usually not killed. He was put into "captivity." What is this captivity? The word "captivity" does not mean prison, as some might think in our modern vernacular. It takes on an interesting connotation in how the Jaredites use it.

According to the dictionary, it could mean prison, but it could also simply mean "confined." But for the Jaredites, it is not prison in the sense we would think of, because the former king has sons and daughters while in captivity. Sometimes many generations of a king's descendants are in captivity.

So, what is this captivity like? This question will be addressed later in this book.

(Note: The word "captivity" was used, for example, to denote the Jews in exile. They also had children and lived somewhat normal lives. They were just excluded from leaving or doing certain things they would like to do, and what they could do was at the king's discretion.)

When the Jaredites captured the opposing king, only in one case was the winner planning on killing the deposed king (Ether 7 – Noah planned to kill Shule, but Shule's sons saved him and killed Noah.) In all other known cases, once the king was captured, there was no indication that the new king planned to kill him. This appears to be true even in cases where there was huge animosity between the leaders, such as when Coriantumr's sons captured Shared and freed their father.

Some may claim that this was because the new king and the captured king were related. This was often true. They might be son and father, or brothers. But there are enough times where no relationship is given to question this as the motivation.

Further, consider the state of the captured king. He lives in "captivity," where he enjoys the blessing of having children, meaning he still has a wife. Sometimes he fathers a significant number of children. Sometimes there are enough for his sons to create their own army unit to avenge their father. This indicates that the captured king probably has multiple wives and is therefore living a life of nobility even in captivity.

In comparing this to our western culture, it is almost hard to know where to start. Imagine a Western Civilization country capturing one of its most ardent enemies. What would that country do with him?

16

Think about what has been done in wars for decades in Western Civilizations. Usually, the leader would be killed or locked in prison. They may even be tortured. The captured leader would not receive gentle treatment. To think of the captured leader being allowed to live some semblance of their previous life of nobility or high society is quite unthinkable.

It hardly matters which Western society a person chooses; examples abound. Think of the leaders captured in the world wars. In the Nuremberg Trials, twelve men were convicted to be hung. Ten were killed, and the other two committed suicide first [2]. Consider more recent conflicts, such as the Gulf War. When Saddam Hussein was captured, he had a trial which some considered a quick, mock trial, and then he was hung [3].

Not only would many armies not treat captured leaders kindly, but many army leaders would not even consider letting the leader from the other side live. This includes armies of organized and supposedly humane countries, as well as armies of organizations considered terrorists, such as ISIS.

Of all the ideas presented in the book of Ether that are strange to us in our Western Civilization, this one, of allowing a captured leader to live in relative comfort and nobility would have to be the most unusual. Of all the skepticism I have heard about the book of Ether from people, this one is the highest on the list. The contrast between the Jaredites allowing the captured leader to live in this manner and the stark realities of how a captured enemy leader would be treated in Western Civilizations is in such opposition that it seems impossible that such a civilization as the Jaredites could exist. It is easy to see why so many were skeptical of the book of Ether.

What did the ordinary people do once they got new leaders?

All indications are that the minute there is a new leader, the ordinary people accept him as their king. Even the military seems to accept him. There is almost no indication of the people ever refusing to accept the new king. What makes this acceptance unusual to us is that their change of allegiance doesn't depend on whether the new leader becomes king by his army defeating the army of the current king, or if the new king is put in place by someone simply killing the current king or putting the current king into captivity.

The ways new kings come about are usually one of three ways:

1. By his army defeating the army of the current king,
2. By sending someone in to kill the current king as a commando-unit-style slaying,
3. The new king kills or captures the previous king himself

For the Jaredites, once the new leader takes over, there does not seem to be any question of his authority by the ordinary people. The people, even the army, seem to turn their allegiance to whoever is the winner. The only ones who seem to defy the new king are those seeking power, not due to a lack of the king's legitimacy, but because of the man's personal ambitions.

The ones seeking power for themselves tend to go into the wilderness and start drawing people to themselves. The draw they offer doesn't seem to be based on grievances the people have with the new king, but more on offers of reward, as seen with the sons of Akish in Ether 9:10-12.

In comparison, in Western Civilizations, the thought that a person would immediately turn their loyalty to the winner of a conflict seems unthinkable. Switching allegiance in an instant is considered the highest treason. Not only would it be condemned in the military, but society, in general, would condemn and ostracize a person who would do this. Words like "traitor" or "turncoat" would follow the person throughout their life.

If the person was part of the army or did something that gave the enemy an advantage, they could be tried for treason and even put to death. The infamy of such people still leaves a scar on them today. This would include such people as Benedict Arnold in the U.S., German collaborators such as Vidkun Quisling of Norway, or U.S. propagandist for Japan in World War II, Tokyo Rose. In more recent times, an example would be James Walker Lindh of the U.S., who joined the Taliban and was considered one of the foremost traitors in the conflict in Afghanistan.

All these stories of traitors, and many more, bring an ugly image to mind. This can be true no matter which side they betrayed. Betraying one's own country and people is considered a cowardice offense in Western Civilization and is seldom accepted for any reason.

The idea that not only would some people accept switching sides of a conflict as acceptable, but that whole countries would do so seemed incredible to the scholars in the day of Joseph Smith. Most people in Western Civilizations would still feel great disbelief today, even though we are becoming more familiar with cultures where that is considered normal and is not believed to be wrong or unethical.

Conclusions of Scholar Skepticism

These cultural differences are much of what laid the basis for some scholars to feel the book of Ether was unusual and therefore have doubts about the Book of Mormon. We must realize that these scholars were mainly only familiar with Western Civilization lines of thought. They had little experience with a culture like the Jaredites. Does that mean cultures like the Jaredites didn't exist? Of course, they did. A person just has to look at the game of chess to see that.

However, the scholars didn't seem to put the two things together, and they may not have been familiar with chess, even though it was around. It wasn't as well known in the United States in the 1800's as in other countries.

In following chapters, we will look at these unusual cultural norms of the Jaredites and why they might not be uncommon for their society. We will consider societies matching the Jaredites and look at archeology and historical information to help us understand them better. Where did they come from? What might they have looked like? We want to learn everything about them that we can.

4
Are There Any Other Cultures That Match the Jaredites?
⊹

For this chapter, to initiate an understanding of cultures like the Jaredites, I want to consider two things that might be familiar to most of us. These items are the game of chess and the 1998 animated Disney® movie *Mulan*. If you haven't seen that movie, you might want to watch it. Even though there is much of Hollywood in the film, many intriguing pieces from the movie are part of the ancient Oriental cultures and match the story of the Jaredites. We can use the cultural norms of these two things to understand the Jaredites better.

Chess

Let's start with the game of chess. In chess everything is done to protect the king, and when the king is killed or captured, the game (war) is over.

This chess scenario played out for me once in high school. I had a math teacher that was a chess master. I was reasonably good at chess and accepted when he challenged me to a game. Because he was so good, his goal was not only to win, but to take every piece I had before putting my king into checkmate.

He had me down to one pawn and my king, and I had only taken about half of his pieces. But in his effort to take all my pieces, he hadn't noticed that I had cornered his king. I put him into checkmate with my king and the one pawn and won the game. That was humiliating to him since it was his first loss to a high school student. (I never played him again, even at his pleading, so I could always remain the champion.)

This same event can be seen repeatedly playing out in the book of Ether. It doesn't matter how many soldiers an army has; if they kill or capture the king, they have won. And it doesn't matter whether the leader is killed or captured. It also doesn't matter whether the leader is deposed by a battle between armies or whether the leader is removed by a coup by what could be considered a commando squad. In all cases, the goal is the same; to control the leader.

Understanding the history of chess can give us insight into a culture or cultures that are similar to the Jaredites.

Figure 2: The Game of Chess

Chess History

According to Murray, Chess seems to have either been invented in India and spread to China or was invented in China and spread to India [4]. Murray leans to India as being the place it was created. There are many legends about it, and we will explore one later. Either way, it appears to have had its roots in Asia. Some scholars feel chess, or some variant of it, was invented as early as 455 BC. It may have gone through some iterations, but most scholars think that by 800 A.D., the game was well known in a form close to what is familiar to us today. That would mean that when Joseph Smith started his work in the 19th century, the game of chess was probably quite familiar to many. However, it was not as common in the United States as in other countries.

Murray says that chess was modeled after the way of fighting in the Orient [4]. The war could end with the killing or capturing of one of the kings, or the destruction of one king's army, so he was left helpless. In the case of the king being killed or captured, his army would then fade away.

Murray states that there are many versions of chess, and many Oriental countries have developed their own variation. I, myself, played one of these. The main thing I remember about this game was that instead of each piece being able to capture the king, the only piece in that game that could was the one that would be equivalent to our pawn. The other pieces could take everything else, but only the pawn could capture the king. My strategy had previously always been to willingly sacrifice the pawns for the sake of winning, but in this game, if you lost all your pawns, you couldn't win. And if both players lost all their pawns, the game was a draw because the king could not be killed or captured. I admit, I lost every game of that form of chess that I played.

This different game raised another question or idea for me. Can only certain people kill or capture the king for it to effectively end the war? References about war between Oriental leaders usually say that one leader defeated the other in battle and slew him. This could mean one army won, then the king of the winning side was given the right to slay or put into captivity the opposing king. When Coriantumr was willing to give the kingdom to Shiz if Shiz would spare the people, Shiz agreed as long as Coriantumr would give himself up so Shiz himself could slay him with his own sword (Ether 15: 4-5).

According to Nibley, the ancient battles were often a chivalric competition between the two kings [5]. They had to be last to fall. He says that the king would be pressed until they could not move, and as has been previously mentioned, there appears to be a preference to capture the opposing king instead of killing him.

The game of chess ends with the phrase, "checkmate." When I was young, I used to think that phrase was English, and when I said to my opponent, "check, mate," I tried to use my best English accent. However, Murray says that "checkmate" comes from "shah – māt." A shaw is a king, and "māt" is Persian and means helpless or defeated [4]. So, the phrase is saying to your opponent, "Your king is helpless or defeated." It is not saying he is to be killed.

Murray claims that the core concept of the six main variations of chess are all the same. Even the pieces are set in the same basic pattern with the king in the center, rooks in corners, knights next to rooks, and bishops next to knights. All have pawns in the front, and all games end with checkmate.

So, why is the history of chess so important here? The answer is that since the game of chess aligns so closely with the wars in the book of Ether, military ideas quite different from our western ones, there is likely a connection between these two cultures. Could the Jaredites and the cultures of the Orient come from the same roots? By studying the cultures of the Orient, we can gain some possible understanding of the Jaredites.

Disney® Movie *Mulan*

In the 1996 Disney® movie *Mulan*, many things are similar to the book of Ether. One of the most important is a statement made near the beginning of the movie that most people miss. In my seminars, I ask the participants what the second scene in the movie is. If you have seen the movie, pause for a moment from reading, and see if you can name it. Have you paused? I will give the answer a little later so you can see if you are right. But right now, let's consider the legend that *Mulan* comes from.

According to ancient Chinese legend, Hua Mulan was a warrior woman [6]. I think the first time I came across this legend and story was while I was studying the history of mathematics. In particular, I was looking at the growth of math in China and other parts of the Orient. It is hard to read too much about China without coming across the legend of Mulan. The Chinese feel she was a real person and a great hero to her country. The stories of Mulan are often in poetic form and include ballads and theatrical works. Figure 3 shows a statue built in her honor.

Figure 3: Hua Mulan

There are a few things that are key in all the different versions of the stories I have read. First, Mulan entered the army in place of her aging father, whom she loved, knowing he could not endure the viciousness of war. Second, she always seems to have hidden the fact that she was a woman until later in the story, and when it did come out, that inspired her people. Third, there is usually a love interest in the form of a military commander. And fourth, she doesn't seek for riches from her fame and power, but only to do good for her people and her family.

Let's go back to the movie version by Disney®. Let's lay out the key concepts it shows us. To keep these concepts separate and distinct, I will number them.

Mulan Concept 1
Protect the King at all Costs

Do you remember what the second scene was? Most of those attending the Education Week class guessed the scene where Mulan is feeding the chickens, but that is scene three. Some even guessed the scene where Mulan is going to the matchmaker. But that is even later. Let's start with scene one. Scene one is when Shan Yu's army comes up on the Great Wall of China, and the fires are lit to tell China he is there.

So, what is scene two? In scene two, General Li goes into the emperor's palace and informs the emperor that Shan Yu's army has crossed the border. Then the general says he will set up a perimeter around the palace to protect the emperor. The emperor, in his goodness, tells the general to protect the people.

Did you catch what General Li was saying? Everything was about protecting the emperor. This is the idea of the game of chess. Everything is set up to protect the king. It was as if the ordinary people were an insignificant detail. But in the movie, the emperor's benevolence shows in his desire to protect his people, even over his own safety. Even so, we see the concept of chess in that scene, to protect the emperor at all costs, even over the common people's lives.

Mulan Concept 2
It Doesn't Take a Whole Army to Win

Another event of interest to us occurs later in the movie. When Mulan drops the avalanche on the Hun army, only four survive. You might think, okay, Mulan is a hero; she goes home, roll the credits, end of the movie, we are happy, we feast. But that is not the case. The four, one of whom is Shan Yu, go into the city to finish their job. All they have to do is kill or capture the emperor to win, and they don't need an entire army to do that. What you see at that point is the idea of a commando unit capturing the emperor or killing him.

Mulan Concept 3
Control of the Emperor is Control of China

You may note that once Shan Yu gets the emperor on the balcony in front of all the people, he then tells the emperor to bow to him, claiming the emperor's walls and cities have fallen to Shan Yu. To us, this may seem like someone who has lost his grip on reality. He is in the midst of a huge city, surrounded by thousands, if not millions, of people. Just below the balcony on which he stands is the emperor's army. This does not sound to us like what he describes, fallen walls and cities. But to Shan Yu, with the emperor standing before him as a prisoner, the statement is very real. If Shan Yu controls the emperor, he controls China.

Mulan Concept 4
Bow in Captivity or Die

Shan Yu, holding the emperor as his prisoner, now commands the emperor to bow to him. Every indication is that if the emperor bows, his life will be spared, and he will be Shan Yu's captive. In this situation, if the emperor does bow and is put in captivity, the people will submit to the authority of Shan Yu.

However, the emperor refuses to bow. It is at this point that Shan Yu is going to kill the emperor. Through the movie, Shan Yu shows total disregard for the lives of soldiers and ordinary people. But when it comes to the emperor's life, Shan Yu is willing to spare his life and take him captive if the emperor will bow. It is only when the emperor will not bow that Shan Yu chooses to take his life. (Spoiler alert for those who haven't seen the movie. Mulan helps save the emperor. I just saved you eight dollars having to see the movie at the movie theater.)

This shows Shan Yu's desire to have the emperor as a captive over killing him. This is very much like the book of Ether, where the Jaredite kings tend to spare the life of the enemy leader, but not the lives of other people. Remember in Ether 8, King Jared's army is destroyed but his life is spared.

Summary of Mulan Movie and Culture Matching the Jaredites

Though there is much of Hollywood in the movie Mulan, in many respects it holds true to the traditional story of Mulan found in books. In that movie, we can see the very essence of the culture of the Jaredites. First, everything is done to protect the king. His is the only life that seems to be of much value. Second, it does not take an entire army to win—a commando squad can win if they can kill or capture the king. Third, control of the king is control of the nation, whether he is dead or alive. And fourth, it is preferable to capture the king and keep him in captivity instead of killing him.

There is much to discuss in these cultural ideas that may seem strange to us. We will return to them further on in the book and discuss plausible reasons for these cultural practices that we find unusual.

5
Time Period and Travel
+

Timeline of the Jaredites

Before we take on the question of customs of the Jaredites we find unusual, we need to consider their time and travel. We can calculate the timeline for the Jaredites based on the Old Testament and the flood of Noah. Noah was probably born somewhere around 3000 to 2900 B.C. This would put the flood somewhere between 2400 and 2300 B.C. [7], though some go as late as 2000 B.C. The tower of Babel was probably a hundred years or so after the flood, and the people had again started populating the earth [8]. Others might put it even two-hundred years after the flood. But even non-LDS scholars tend to put it in this time frame [9]. So, it would probably put the tower of Babel somewhere between 2500 and 2000 B.C.

One challenge in timelines that you often see is that it is hard to grasp the proper perspective of the time length of the Jaredites compared to the Nephites, Lamanites, and Mulekites. I created a time perspective timeline using 2350 for the time of the flood, 2250 for the time of the Tower of Babel, and the information from The Book of Mormon. The lengths were done by using pixels to measure the intervals. The only exception was Lehi's departure from Jerusalem and that of the Mulekites. They were only eleven years apart, and due to the timeline's length, eleven years would make them almost right on top of each other. I put them as close together as I could. This perspective can be seen in Figure 4. I also did it vertically since there is more room on the page to give the perspective.

```
2350 BC      | Flood
About 2250 BC | Tower of Babel - Jaredites




600 BC       | Lehi Left Jerusalem
589 BC       | Mulekites Leave Jerusalem


130 BC       | Nephites & Mulekites Join
33 AD        | Christ Comes
About 201 AD | Wickedness Starts Again

About 421 AD | Moroni Hides Plates
```

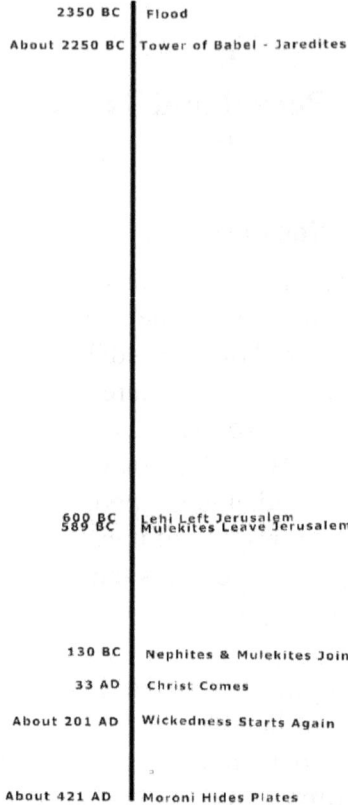

Figure 4: Time Perspective Timeline

The first realization a person should gain from this is that the Jaredites lived in the western hemisphere far longer than the Nephites. If they crossed the ocean around 2250 BC and survived until the Mulekites found Coriantumr, this was likely more than 1600 years. On the other hand, the Nephites were in the Americas for around a thousand years.

The Jaredite time on the American continent could have even been substantially longer than 1600 years for two reasons. First, they could have come earlier than we are estimating if the flood and Tower of Babel were at the earlier end of the suggested time frame. The second reason is that we don't know how long the Jaredite culture overlapped with the Mulekites and Nephites. It could have been from a few years to decades to possibly even a century or more. The Nephite and Mulekite nations existed on the same continent for around 450 years before Mormon brought the Mulekites into the story. The fact that the end of the time of the Jaredites was after the Mulekites' arrival comes from only three short scriptures in Omni.

Omni 20 And it came to pass in the days of Mosiah, there was a large stone brought unto him with engravings on it; and he did interpret the engravings by the gift and power of God.

Omni 21 And they gave an account of one Coriantumr, and the slain of his people. And Coriantumr was discovered by the people of Zarahemla; and he dwelt with them for the space of nine moons.

Omni 22 It also spake a few words concerning his fathers. And his first parents came out from the tower, at the time the Lord confounded the language of the people; and the severity of the Lord fell upon them according to his judgments, which are just; and their bones lay scattered in the land northward.

Thus, the Jaredites could have existed as a nation for up to 450 years after the Mulekites came to America based on Mosiah's reign being the only certainty we have on the Jaredite timeline from the Mulekites. I have reasons to doubt that it was more than about fifty years after the Mulekites arrived before they found Coriantumr. I also want to be clear that I am not saying that the different nations did not cross paths previous to them finding Coriantumr. As I will show later, they likely did. But it appears that Mormon is trying to keep his story Nephite-centric. Tangents would likely lead away from his message and the lessons God is inspiring him to teach us.

The key point is, the two bookends of the Jaredite's existence, the Tower of Babel, and the finding of Coriantumr by the Mulekites, give us some idea of the length of the Jaredite nation's existence. However, because of the uncertainty of the exact years of these two events, this range could be hundreds of years more or less than the 1600 years approximated here. But, for sure, that existence was probably more than one and a half millennia, a long time for a nation to exist. To put it into perspective, that's around eight times longer than the existence of the United States at the time of the writing of this book.

There is another interesting way to look at the length of time the nation existed. Thor Heyerdahl was one of the most famous archeologists that ever lived. One of the things that made him good at historical investigation was that he didn't just try to figure things out by looking at archeological facts, but he would also gather stories, legends, and history as much as possible from the people in the area.

For his work in South America, he mentions that there was a brown-man nation and a white-man nation that were constantly at war with each other. This, of course, is very familiar to those who have read The Book of Mormon. However, Heyerdahl was not a member of The Church of Jesus Christ of Latter-Day Saints and was not relying on The Book of Mormon. He collected stories from the people who talked about how the brown man eventually destroyed the white man. From the legends of the people, he recorded the names of the chiefs back to that event, then multiplied them by 25, the amount of time he felt would be an average for the amount of time each chief would serve in that capacity. By doing this, he determined that this great destruction of the white man occurred around 500 A.D. [10].

This method is amazingly close to what The Book of Mormon says. But there are some challenges in using this method in the book of Ether. First, what lineage we have is not just of the kings, but follows the genealogy leading to Ether, the writer. Also, there could be gaps. In Ether 1, most of the lineage says that a man is "the son of" the one before. But a few say that the man was "a descendant of" the person ahead of them. This could mean there were multiple generations in between. This occurs three times. They are listed here, along with information about the same men that comes later in the book of Ether.

Table 1 - Descendants

Man/Descendant of	Related Information Later in Ether
Ether 1:6 . . . He that wrote this record was Ether, and he was a **descendant of Coriantor**	Ether 11:23 And it came to pass that **Coriantor begat Ether**, and he died, having dwelt in captivity all his days.
Ether 1:16 And Aaron was a **descendant of Heth**. . .	Ether 10: 31 And he begat Heth, and Heth lived in captivity all his days. And **Heth begat Aaron**. . .
Ether 1:23 And Morianton was a **descendant of Riplakish**.	Ether 10: 9 And it came to pass after the space of many years, Morianton, (**he being a descendant of Riplakish**) gathered together an army of outcasts, and went forth and gave battle unto the people; and he gained power over many cities; . . .

Twice when it says descendant it later says begat. Though the word begat usually means the person is a child of that man, according to Answers in Genesis, it sometimes might take on the meaning of a descendant after multiple generations [11]. But for Morianton being a descendant of Riplakish, it uses the same phrase of descendant both times in the book. It also uses the phrase "after the space of many years…"

Riplakish had been a wicked king, forcing his people to pay heavy taxes, imprisoning and forcing people to work who could not afford to pay the taxes. The people hated him, eventually killing him and driving his descendants out of the land. Morianton, one of his descendants, came back later. The scriptures say it was many years later (Ether 10:4-12). It could even have been generations later.

The main point is, we do not know for sure the lineage from father to son. We likely have a gap in the lineage record with Riplakish and Morianton, and possibly with the other two as well. There are 30 men listed in Ether 1, and if our estimate of 1600 holds for these men, that would be over 53 years for each generation. There is usually a lot of overlap between generations. Answers in Genesis says that about 40 years defines a generation [11]. At 40 years per generation, for 30 men, that would suggest a time length of 1200 years. If our assumption of a minimum of 1600 years is correct, that would likely mean some in the generation list are missing. This gives us three options to consider. Either some are missing from the generation list, the Jaredite nation existed for less than 1600 years, or there were more than 40 years per generation. The most likely option is that men are missing from the lineage list. Ludlow [12] also considers this to be a strong possibility.

In final, what this means is that trying to use the lineage list as a means of determining the length of time the Jaredites existed as a nation would likely not be an accurate measure.

With the understanding of the timeline for the Jaredite nation, let's now turn to the second and third parts of the question, from where did they start, and where did they travel?

Where Did the Jaredites Start From?

We have talked about unusual parts of the Jaredite culture and how it relates to the game of chess and the story of Mulan. The question is, could the similarities in the Jaredite culture and that of the Orient suggest a travel route, and could a travel route also tie them to the similarities in the two cultures? Also, there is an even more important question. Does the book of Ether give us any scriptural clues about the travel?

Can the fact that the game of chess is so close to the ideas of the battles in the book of Ether indicate that the Jaredites had a connection to the Orient?

Let's begin with the land where the Jaredites started their journey. As a joke, I asked Google where the Tower of Babel was. Google came back with an address. It said the address was Iraq. When I told it to map it, it took me to a site some have suggested for the Tower of Babel, and it even had an aerial photo. This site is in a place called Etemenanki. Figure 5 is a picture of the ariel view of the spot many believe was the place the Tower of Babel was built. Figure 6, is a google map of Etemanki and shows it right near Babylon. Finally, Figure 7 is an expanded map of the Middle East showing Etemanki in relation to modern countries and cities.

Figure 5: Etemenanki - Possible site of Tower of Babel

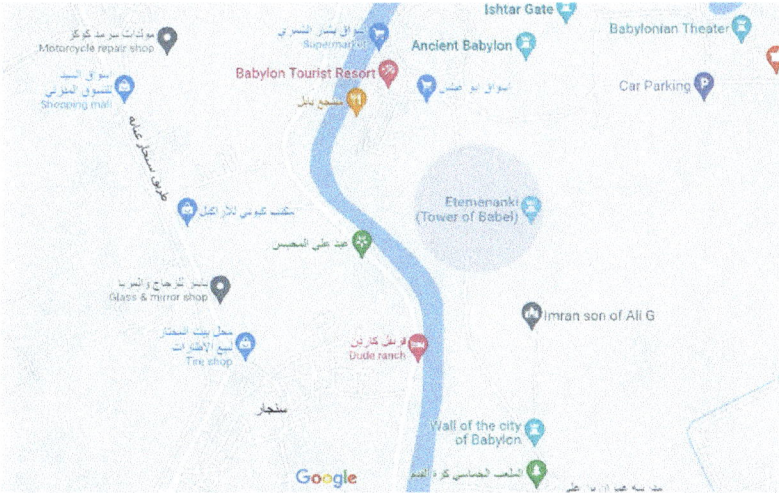

Figure 6: Etemenanki - Assumed site of Tower of Babel

Figure 7: Etemenanki map of Middle East

One of the reasons this site is considered the place of the Tower of Babel is because of writings and images cut into a rock. The rock appears to show a figure of a great king and a tower. Figure 8 shows the rock, and Figure 9 shows it with the image highlighted in yellow.

Figure 8: Image at Etemenaki assumed to indicate Tower of Babel

Figure 9: Rock carving at Etemenanki highlighted

Of course, we don't know exactly where the tower of Babel was. The Bible tells us it was in the plains of Shinar. (Genesis 11:2).

According to Answers in Genesis [13], there is a place named Shinar. However, it is unlikely that is what the Bible is referring to since it didn't exist as a place when Hebrew was a local language. Shinar appears to have different spellings and different variations in many of the cultures and languages of that area. It also has quite a few interpretations.

Habermehl [13] says some of these interpretations include "to shake out" as a reference to the dispersion of the people from the Tower of Babel. But Habermehl says that he believes that Shinar is instead a combination of two Hebrew words, "shene nahar," which would be "two rivers." This would mean the city was where two rivers met, diverged, or were near each other.

Habermehl [13] goes on to say that this would be close to the Greek term, "Mesopotamia," meaning the land of the two rivers. This is a name often used to reference the land around the Tigris and Euphrates rivers. This would be the area between modern-day Iraq and Iran.

This area would agree with most accounts, and Etemananki is in this general area. The Tower of Babel would have its roots in what would become Babylon, or the Mesopotamia Valley. Though the word Babel and the word Babylon are not necessarily related, Habermehl [13] says that different translations of the Bible sometimes interchange Babylon and Shinar. But by all indications, the Tower of Babel was likely somewhere in the land of the Tigris and Euphrates rivers.

A side note might be of value here. Christian and Jewish tradition has it that the king responsible for building the tower was named Nimrod [14]. Sometimes he is called the king of Kish, a Sumerian city-state also often associated with the Tower of Babel and the Mesopotamia area.

Genesis 10:9-10 says that Nimrod, the mighty hunter, had Babel as part of his kingdom.

Genesis 10:9 He was a mighty hunter before the LORD: wherefore it is said, Even as Nimrod the mighty hunter before the LORD.
Genesis 10:10 And the beginning of his kingdom was Babel, and Erech, and Accad, and Calneh, in the land of Shinar.

I've read the works of many people who suggest it was Nimrod who instigated the work on the Tower of Babel. Hugh Nibley is one who says that [5]. Nibley also says that Nimrod is the name of a legendary man called the Mad Hunter of the Steppes. One thing for sure is that the name Nimrod is common among the Jaredites both as a place name and later as the name of one of the Jaredite kings.

Let's now return to the travel of the Jaredites. Consider the following map in Figure 10 of the world as we currently know it. I have indicated an arrow to the area most of the scholarly world assumes the Tower of Babel could have been, that of Etemenanki. From there, we will consider the question of where the Jaredites traveled when they left.

Figure 10: Map of Middle East with approx. Tower of Babel location

Where and How Did the Jaredites Travel?

We can break this question into a few parts. First, after Jared and his brother had gathered their families and friends and their friends' families, which way did they go when they left the Tower of Babel in the plains of Shinar? Second, where did they travel? And third, how did they travel?

If you refer to the map above, and I asked where you thought the Jaredites would go, what would you answer? In my seminars, I ask those who know what the book of Ether says to not answer. I then invite those who are unsure to guess. The most common guess is west. The second most common guess is south, and the third is east. But, in fact, the real answer is north.

Ether 2:1 And it came to pass that Jared and his brother, and their families, and also the friends of Jared and his brother and their families, went down into the valley which was northward, (and the name of the valley was Nimrod, being called after the mighty hunter) with their flocks which they had gathered together, male and female, of every kind.

The reason it seems strange for them to go north is because that is not the way you would expect them to go if they are heading to America. East or west seems reasonable. Even south, as Lehi did, would get them to a place to travel to America. But north? What is there northward to take them closer to America?

Hugh Nibley said the Jaredites likely went north and then traveled across the Asian steppes to modern-day China [5]. That would be the Siberian parts of Russia today. Much of his reason for this claim was because of the similarities between the Jaredites' culture and that of the Mongols and Chinese.

But when I was younger and read Hugh Nibley's claims, I also read some articles by scholars from The Church of Jesus Christ of Latter-Day Saints who said that they felt it would be impossible for the group of travelers to have crossed that inhospitable land. Still, for me, there was reason to believe that they did make this journey. One was a key scripture in Ether.

Ether 2:5 And it came to pass that the Lord commanded them that they should go forth into the wilderness, yea, into that quarter where there never had man been. And it came to pass that the Lord did go before them, and did talk with them as he stood in a cloud, and gave directions whither they should travel.

If they traveled across the great steppes of Asia, it would match what the scriptures say about going into the wilderness, into a quarter where men had not been. But I could also see the concern some scholars had about them being able to survive. Then one day, as I pondered this, I had an interesting experience.

I was at a math conference and was attending a session on the math of global warming, which was becoming a hot topic at that time in the 1980s. (No pun intended. Well, maybe a little.) I had been thinking about the idea of the Jaredites traveling across Asia and was not paying much attention to the speaker. Suddenly, I felt a strong prompting from the spirit telling me I should listen to what the speaker was saying.

The speaker had a slide of Asia on the overhead projector. He said that one of the greatest signs of global warming was the gradual disappearance of what he called The Great Siberian Sea. He said that the sea had reached its peak at about 3,500 to 2,000 B.C., but now (being the 1980s), the last remnants of it were fading away.

I looked at that slide and realized pieces of the puzzled I had not understood. The land the Jaredites knew was not the same as the one we know. In fact, being only a hundred years or so after the flood, it would have been far different. Many lands would likely have had water still receding from them. And though the scholar giving the talk did not mention the flood at the time of Noah, which many scholars would have laughed at, the thought was clear to me.

The flood of Noah was estimated in the time period he said, of 3,500 to 2,000 B.C. That would make sense why The Great Siberian Sea was at its highest point then. Chances are that if it was just fading away in the 1980s, Siberia in the time of the Jaredites could still have a lot of water covering it.

As I put this book together, I tried to find any image that matched the one in the slides the scholar showed that day. But information on The Great Siberian Sea, as he described it, is minimal, and I have yet to find a map. So, I devised a plan to replicate his slide the best I could. I took an elevation map of central Asia and replaced the lowest elevations within the mountain ranges with blue. The original map can be seen in Figure 11, and the revised map is in Figure 12.

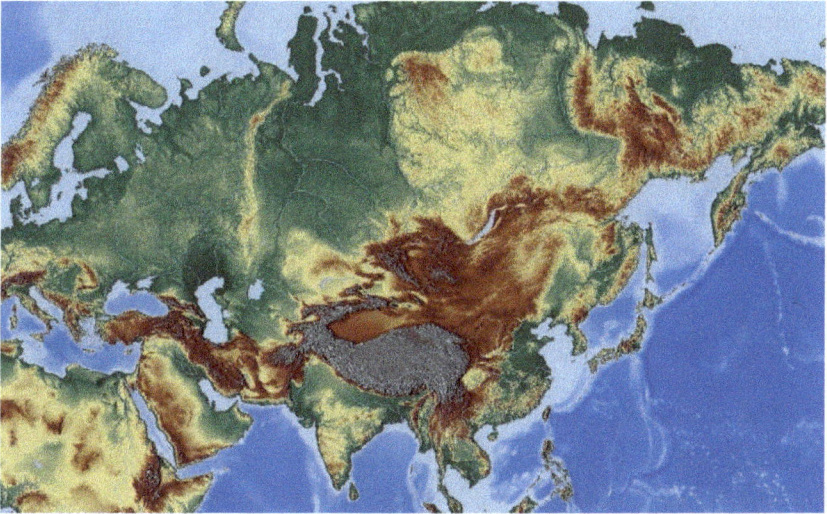

Figure 11: Relief map of Asia

Figure 12: Relief map of Asia (Edited)

The map is by no means a perfect rendition, but it will give the basic idea of what the slide looked like that day. I have also marked the approximate place of the Tower of Babel in the Mesopotamia valley with an arrow in Figure 12. If the Jaredites went very far north at all, they would likely have run into The Great Siberian Sea.

One thing to note here is that the Great Siberian Sea the man showed on the map, which I tried to duplicate, was at its highest level.

The Jaredites probably came hundreds of years later, and this sea had likely shrunk some. But it probably still had to be a significant body of water if it took until the 1980s for it to totally disappear. However, there would likely have been more land around its perimeter at the time of the Jaredites than this map shows.

At the time I learned about the Great Siberian Sea at the math conference, I hadn't heard about it from Nibley. He does mention it in his book [5], which came out at about the same time I learned of the sea in the seminar. I don't know if he had already written about it before his book came out, and I had missed it, but it was good to see we had some agreement there.

The scriptures give us another clue that this sea was probably there, and that is how the Jaredites traveled after going north. Ether 2, right after saying that they were led into the wilderness into a quarter where men had not been, says the following:

Ether 2:6 And it came to pass that they did travel in the wilderness, and did **build barges**, in which they did cross many waters, being directed continually by the hand of the Lord.
Ether 2:7 And the Lord would not suffer that they should stop beyond the **sea in the wilderness**, but he would that they should come forth even unto the land of promise, which was choice above all other lands, which the Lord God had preserved for a righteous people.

It is interesting to note a couple of things here. They built barges, and it appears they made some type of craft that was familiar to them. They did not have to be schooled in how to build them as the brother of Jared later was by the Lord when they made the ships that were "tight like a dish." So, what kind of vessel were these barges?

The archeologist, Thor Heyerdahl, believed ancient cultures did not build wooden ships like we are familiar with in the times of the Romans and Greeks. He felt that the vessels that were built when men first started water travel were of two types—rafts made from logs, and reed boats. To prove his theory that ancient civilizations could travel such distances on these types of barges, he traveled across oceans in such crafts. The Kontiki, his first such ship, was made of balsa logs [10] on which he and his crew crossed the Pacific. Later, he and his team crossed the Atlantic Ocean on a ship made of reeds [15].

In crossing these oceans, Heyerdahl showed it was plausible that man-made rafts and boats built out of materials like logs and reeds that were familiar to men anciently, could make such journeys. It is possible that the Jaredites crossed the great Siberian Sea using one of these methods. Figure 13 shows a picture of Thor Heyerdahl's log raft Kontiki on which he and his crew crossed the Pacific Ocean. Figure 14 shows a picture of Ra II, the reed boat on which Herydahl and his crew crossed the Atlantic.

Figure 13: Heyerdahl's Kontiki

Figure 14: Heyerdahl's Ra II

It is worth noting that Heyerdahl had built a previous reed boat that was named Ra. It was his first attempt to cross the Atlantic in a reed boat, and he failed. The boat was about a hundred miles from the Caribbean islands when it started to break apart, and they had to be rescued. However, for Ra II, Heyerdahl hired some men from Peru, where they built boats on Lake Titicaca. With their increased skill over the builders who used ancient Egyptian images as their guide, Ra II made it across the ocean with no mishap [15].

I had my own encounter with a reed boat on Lake Titicaca. I had the chance to join the faculty from Brigham Young University - Idaho for an excursion to Peru. We visited many places, including Nazca, Machu Picchu, Cusco, and many other areas. When we were at Lake Titicaca, we took a motorboat out to visit the floating islands.

The books about Peru said that visiting the floating islands was like stepping back in time hundreds of years. I had always thought the floating islands dated way back to ancient times. However, as I studied about them, I found out that the South American natives created them to escape into the lake to avoid the Spaniard's persecution and enslavement. Still, to step onto the islands, back to a time of centuries previous, was a fascinating opportunity.

We chartered a boat to make the trip out to some of the islands. Because of the altitude, many in our group could not make it on the boat and instead had to stay in the hotel near the oxygen.

We scheduled to visit three islands. When our boat landed at the first one, and I stepped from the boat onto the island made of reeds, it gave me the sensation of stepping onto a firm waterbed. The reeds gave some under my weight but quickly bounced back as I moved on. See Figure 15.

Figure 15: Pulling up to a floating reed island.

And even though we were supposed to be stepping back to a time centuries earlier, the little Peruvian women were there with their tables selling modern trinkets, though they might have done that anciently as well.

I have always been fascinated with how people live, so while my colleagues bought necklaces, trinkets, and small bags with llama pictures sewn on the sides, I went to one of three reed huts on the island. Using more signing than language, I asked the man if I could see what his home was like.

He smiled and said the usual reply, "One sol or one dolla'." The exchange rate at the time was three sol to a dollar, so I made sure I carried a good supply of sol for just such purposes. After all, a person doesn't want to lose their sol (soul). So, I paid him the sol, and he moved aside the colorful tapestry that hung over the door to escort me into the one-room hut.

The room was about ten feet by ten feet (3.5 meters by 3.5 meters). The furniture inside was all made of reeds. There was a reed bed that was also meant to be used as a couch. There was a small reed table. And there was a reed piece of furniture about three feet (1 meter) high, three feet (1 meter) wide, and two feet (.6 meters) deep, like a cabinet. To my surprise, a television was sitting on this last piece of furniture. It was on with the sound turned down, and a soap opera was playing.

I must have had a shocked look on my face because the man laughed. I had, after all, read that stepping on one of these islands was like stepping back hundreds of years, and though I may be wrong, I was sure there were no televisions hundreds of years ago. Also, when a wind came up, it blew the islands around the lake. Our tour guide said he had been a mailman delivering mail once per week to the different islands on the lake. He said that he would check the past wind directions that had blown the previous week before he left for the delivery. That way, if he got to the place an island had been, and it was not there, he would know which way to go. So, how could they have something as modern as television in that environment?

To my inquiry about the television, mostly done by pointing and questioning gestures, the man said, "Me show." So, when I finished the tour of the inside of his hut, we went outside. He put his hand on my shoulder and pointed up to the roof.

At first, I couldn't see what he was pointing at, but considering I had just seen a working television, I knew it had to do something with that. I looked carefully, and I realized what was there. He led me around behind his hut and pointed at the roof where I could get a better view. Hidden by reeds on the front, though clearly visible in the back, was a solar panel (Figure 16). When I realized what he was telling me, I turned to him and nodded. He then smiled and said, "This year, TV. Next year, internet."

Figure 16: Solar panel on the roof of Peruvian hut on floating island

He and I visited for a while using lots of gestures, and him using his limited English and me using my almost non-existent Spanish. When I could tell my colleagues were finishing their purchases, I returned to my group. But just before we were ready to board our boat for the trip to the next island, a reed boat pulled up (Figure 17).

It wasn't a big boat, but it did have an upper deck. I marveled at the weaving of the reeds on it, and I hoped I could pay to tour it.

Figure 17: Reed boats on Lake Titicaca

Our guide interpreted for us and said the men were willing to take us to the other islands on our itinerary. It was just ten dollars per person. Most of the group started questioning whether such a boat would float, but I was eager to try it. I was first in line to pay my ten dollars, afraid it would fill up. But I needn't have worried. Only about five of us took the opportunity. Our tour guide told us it was perfectly safe, but our group leader had his doubts. Just for safe measure, we retrieved our life jackets from our motorboat and put them on.

Since the men would row us across the lake instead of motoring like our modern boat, they suggested we get a head start. As the five of us loaded up, we had to endure the teasing from the others in our group.

"See you at the bottom," someone yelled.

"Hope you make it back to the mainland sometime before next week when we leave," another called out.

It seemed like everyone had to add their two bits of wisdom, though for some it was quite a bit less in value than two bits. Many of them said they wouldn't trust anything that didn't use modern technology. Ignoring all their comments, the five of us on the boat soon enjoyed the smooth forward strokes of the Peruvians rowing us across the lake.

I climbed on the upper deck so I could enjoy the view. For most of the trip, I had it all to myself. The others were a little nervous about it tipping and dumping them off. But the boat was extremely steady and didn't rock nearly as much as the motorboat did that we had come in. It was a very pleasant trip. The men told us the boat, though sturdy, was relatively light. They could row it at a good pace.

**Figure 18: Traveling on reed boat on Lake Titicaca
(author is on top deck)**

When we got to the next island, we assumed our group would already be there, but they weren't. The three of us received a royal welcome and a grand tour all to ourselves. We wondered if our motorboat had come and left quickly, but the people said no other boat had arrived. However, since none of us spoke much Spanish, and they spoke little English, we weren't sure whether they actually understood what we asked or if we misunderstood their answer.

After we had been there for quite a while, past when our motorboat should be going to the third island, we decided we should leave. Perhaps the boat had come and gone, and we would be left behind.

We again boarded the reed boat and enjoyed the pleasant trip across the lake. There was no smell of gasoline fumes from a motor, and a gentle breeze blew across the lake. The sky was clear, and it was a beautiful day.

When we reached the third island, again our boat was not there. Once more, the Peruvians greeted us like royalty, and we enjoyed our visit with them. We bought a few things and waited for our boat to arrive.

As it grew late, an hour or more past the time we were supposed to be heading back to the mainland, we became concerned that our tour guide, and the men rowing our boat, might not have adequately

48

communicated which additional two islands our group was supposed to visit. We finally started to see if we could find a way back to the mainland. (The men and reed boat we had come on had already left for their homes.) Suddenly, we heard a motorboat approaching. Sure enough, it was our boat.

When they arrived, we asked them where they had been. In disgust, one of the group told us they had motor problems and had been stranded much of the day out in the middle of the lake.

They sported bright red sunburns for their day due to the sun reflecting off the lake. So much for modern technology.

Now, back to the Jaredites. It appears they built "barges" in a way that they were familiar with because the Lord mentions that later when they built the barges that are tight like a dish, they had to be taught. A barge, by definition, is a flat-bottomed boat, perhaps like the one I rode on. The flat bottom is probably why it felt more stable than the modern-day boat.

The critical point of the story diversion is that the little reed boat I rode on made me very comfortable with how well-built such a craft could be. That might be the kind of boat the Jaredites used, since it was one of the two types Thor Heyerdahl felt people of that era built. I feel it is more likely that they used, the other type, log rafts to cross the "sea in the wilderness."

Heyerdahl made it across the mighty Pacific Ocean on a log raft in 101 days [10] and across the Atlantic Ocean in a reed boat in 57 days [15], and he felt they were safe vessels to travel on when appropriately built. The first reed boat did fail, but when he had the second done by those who were more familiar with creating reed crafts, it made the voyage easily with no sign of wear to the boat.

So, if the Jaredites built either of these types of barges, no one should have doubted that they could travel across the Great Siberian Sea. Perhaps they were even more advanced than we can imagine and built something even better. Nibley [5] says ancient people may have been far more advanced than we think. He says evidence shows no continual rise in advancement, but that knowledge and development rose and fell and rose again. We do not know where technological development was at the time of the Jaredites.

There is one last point to add here. The prevailing winds in Europe and Asia are from the west to the east (westerlies), as seen in

Figure 19. (Arrows at the top of the image). This means that if the Jaredites launched barges into the Great Siberian Sea, the direction of the winds would have pushed them toward China.

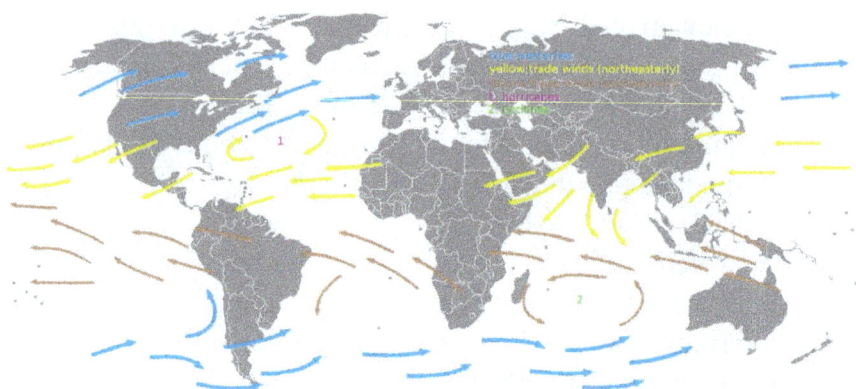

Figure 19: Trade winds of the world

To continue, I'm going to assume the Jaredites followed this route. You may disagree with this, and that is okay. It may give you some ideas you can use to formulate your own opinions. But if they did go north, built barges (Ether 2:6), and crossed some inland sea, they probably had to go east since the prevailing winds in that part of the northern hemisphere are westerlies, blowing eastward.

Something else points to the China coast for the Jaredites' departure across the ocean to America. Nibley [5] reminds us that the mountain where the brother of Jared had the Lord touch the stones was called Shelem because of its exceeding height (Ether 3:1), and there are not any exceedingly high mountains in the direction of the Atlantic seaboard. But there are high mountains on the China coast.

A good example of this is the Huangshan mountains, Figure 20 and Figure 21. In Figure 22, a map, I added the location of the Huangshan mountains.

There are no mountains as tall near the Mediterranean Sea, on the European coast, or on the Arabian Peninsula, other sites that make sense for the Jaredites to depart on their journey to America.

Figure 20: Huangshan Mountains

Figure 21: Huangshan Mountains

Figure 22: Approximate Location of Huanshan Mountains

So, the cultures of China match the story of the Jaredites, and the land matches the description. There is something else that was interesting to me. If the Jaredites were to launch into the Pacific Ocean starting at the China Sea, their boats that were built like a dish had no sails. So even though it says the Lord blew them continually toward the promised land, they would still be very much at the mercy of the ocean current.

I studied the Pacific oceanic currents while reading about Thor Heyerdahl's travel across the Pacific on the Kontiki. Figure 23 shows a map of the North Pacific Gyre, the ocean currents of the North Pacific.

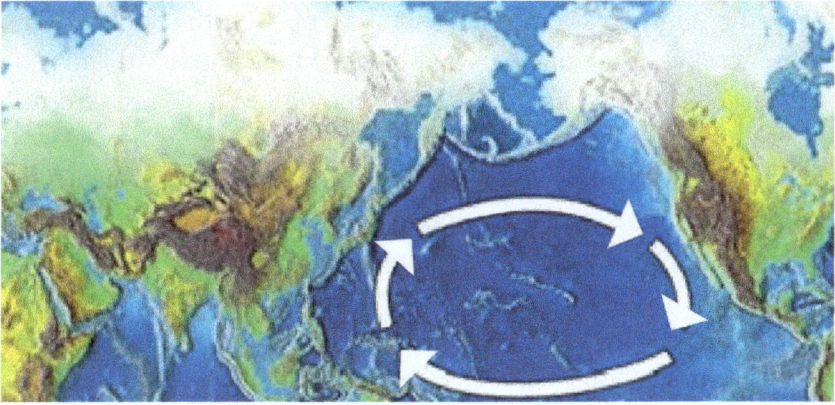

Figure 23: North Pacific Gyre

If the Jaredites launched in tight-like-a-dish barges into the China Sea, the current would have taken them up along the Arctic region and then down along the American coastline. While studying this current, I found some maps with approximate current speeds. (There are different speeds at different intervals of the currents.) Using the listed rates of these currents, along with the distance of each, I estimated the length of time it would take to traverse each current segment. When I added all these times together, it was just about one year for an ocean current to travel from the China Sea to the region around the Canadian-U.S. border. Ether tells us how long they were on the ocean.

Ether 6:11 And thus they were driven forth, three hundred and forty and four days upon the water.

This is shorter than my calculation. Of course, if they followed this route, we cannot be sure where they landed. It could be farther north or south, but Ether also says:

Ether 6:8 And it came to pass that the wind did never cease to blow towards the promised land while they were upon the waters; and thus they were driven forth before the wind.

This means that even though they didn't have sails, the wind could have helped them travel faster than the standard current speed. Still, the similarity of my calculation gives reason to believe this could be their travel route. However, because my analysis could be off, I sought further confirmation of the time traveling on the current. That is when I remembered the tsunami that occurred in March 2011 in Japan. Just over a year later, garbage that had been pulled out to sea from Japan started washing up on the north American coast [16].

Though we do not know the route the Jaredites traveled for sure, there is much evidence for the route proposed. The Jaredites went north from the Tower of Babel and built barges crossing some inland sea. The Great Siberian Sea likely covered much of Asia at the time, providing a route matching this description. The place they built boats to come to America had at least one exceedingly high mountain. High mountains exist along the coast of China, while few other feasible routes do. If they did build their ships near the China Sea and traversed the northern Pacific currents to America, the time travel on the currents matches quite closely to the time Ether says it took them to reach the new continent. And finally, the most significant evidence is that the Jaredite culture matches that of the cultures of the Orient.

Of course, I reiterate that we don't know for sure. And you might have other ideas. That is fine, and I hope you will pursue your own considerations. While searching for understanding, I have felt much direction from the spirit, most directing growth in my life. Quite often, as I came across different ideas in my study, the spirit used them to help me understand what I needed to learn for my own spiritual development. And even if you have other thoughts about what you consider the route of the Jaredites' travel, hopefully, this has given you some interesting concepts to think about.

6
Potential Reasons for Cultural Jaredite Cultural Traditions
✦

In this chapter, we want to look at possible reasons for some of the cultural traditions of the Jaredites that we find unusual. We will do this by considering the culture closest to theirs, that of the Orient. There are many writings from the Orient, but understanding them from our perspective can still be challenging. Culture is passed down through the ages, and sometimes people don't even know why they do certain things.

An example is a story about a lady who always cut the end off of her pot roast before she cooked it. One day her husband asked her why she did that. She considered it for a minute and said she didn't know other than that her mother did it. The husband then called his mother-in-law and asked her why she cut the end of her pot roast. She said she didn't know why other than that her mother did it. He finally called his wife's aged grandmother and asked her why she cut the end off the pot roast before she cooked it. The grandmother said her pan was too small to hold the whole pot roast.

This story shows two things. First, most cultural norms have roots in a valid reason. Second, the reason might have been lost over time. Still, it can be of value to consider the probable reasons why the traditions exist. So, let's start.

Unusual Warfare Customs

Figure 24: Warfare - Genghis Khan Exhibit Museum

Why does the war end when one leader is killed or captured?

First, we should note that even though it is a strong precedent in the book of Ether that the war ends when the leader of one side is killed or captured, that is not always the case. The strongest contradiction to this occurs during the final battles. Coriantumr faces many opponents, with a new leader picking up after the previous leader was killed. However, even then, it still often followed the same pattern, with the war ending at that point, then picking up later after the beaten leader had time to raise his own army. For example:

Ether 13:30 And Coriantumr gave Shared battle again in the valley of Gilgal, in which he beat Shared and slew him.

Ether 14:3 And now, after the space of two years, and after the death of Shared, behold, there arose the brother of Shared and he gave battle unto Coriantumr, in which Coriantumr did beat him and did pursue him to the wilderness of Akish.

This seemed to be much of the pattern that the war ended with the leader being killed, captured, or driven to the wilderness (as Coriantumr was at one point). The only real exception where the fighting didn't stop at the leader's death was with the death of Lib where his brother, Shiz, stepped right in.

Ether 14:16 And when he had come to the plains of Agosh he gave battle unto Lib, and he smote upon him until he died; nevertheless, the brother of Lib did come against Coriantumr in the stead thereof, and the battle became exceedingly sore, in the which Coriantumr fled again before the army of the brother of Lib.

Ether 14:17 Now the name of the brother of Lib was called Shiz. And it came to pass that Shiz pursued after Coriantumr, and he did overthrow many cities, and he did slay both women and children, and he did burn the cities.

So why would the war end when the leader of one side was killed or captured? And why would the armies fight to their destruction to protect their leader? There appear to be three related reasons for these two questions.

War Considered Chivalry Contest Between Two Leaders

The first reason armies would fight to the annihilation of one or both of them was how the contest was viewed. Nibley [5] says that in the ways of the Orient, the war was seen as a chivalry contest between the two leaders. It could not end until one of the two was killed or captured.

At times, the rivalry came down to just the two leaders fighting instead of the armies. Nibley [5] says every war was seen as a personal fight between the two leaders, and sometimes they alone would fight with the loser's people submitting to the winner. However, when instead the entire army came to battle, it often required the destruction of one army or the other. It is almost like a duel in our modern times, but instead of matched weapons that the enemies fought with, they used matching armies.

Often, when one army was destroyed, the leader of that army was brought before the other leader, and if the winner chose not to put him into captivity, he would likely take the life of his opponent with his own sword. A previously mentioned example is when Shiz required that Coriantumr give himself up for Shiz to kill for the war to end. (Ether 15:5). This angered those loyal to Coriantumr, and they continued to fight.

King Divinely Appointed

The second reason armies would fight to the end is that most of the people considered the king to have been installed by God, and that through him many of the blessings of God would come to man [17].

"For nearly three thousand years in China, until the abolition of the monarchy in 1911, the emperor was the focus of those ritual activities by which the prosperity and well-being of the whole empire were established from year to year." [17](p1).

It should be noted that the people did not view the king as a god, as was the case with the Egyptian Pharaohs; at least, the people didn't while the king was still living. However, the king was considered the one whom God had put in charge as the key for the desired blessings from heaven to come. This included blessings of prosperity, health, fertility, life, death, and the seasons. Since China was an agrarian society, depending heavily on agriculture, the blessings from heaven were important. In fact, the king was often given the title "Son of Heaven" as a reference to his divine favor [17].

This divine favor was considered to give the king abilities beyond that of other men, but still did not equate him to the gods. However, he was often considered to have attained a quasi-divine status after he died. He would become the object of worship and sacrifice in the ancient cult of ancestors [17].

Because the people considered the king to be divinely appointed, an offense against the king could be regarded as an offense against God, who had chosen to install the king into his position. This is true even if the king came into power because of war. It was believed that the king won the war because God wanted him to win.

I recently attended a natural history museum that brought in a large exhibit about Genghis Khan. There was a lot of fascinating information, and throughout the exhibit was the underlying theme of the assumed divine intervention in Genghis Khan's behalf. Genghis Khan was the name chosen by him, and it means "universal ruler" [18]. In fact, the exhibit said that Genghis Khan felt God inspired him. Genghis Khan said he knew when he needed to stand and fight and when he needed to retreat. He felt this was especially true when the odds were against him. He claimed this was the key to his success and long life.

Besides feeling the king was divinely appointed, Smith [17] tells us that the king was considered the one who was chosen to set the social order, and it was unheard of not to have a king. Without a king, they assumed societal chaos would ensue. The social order was important in their societies. The more the king did for the good of society, the more the people felt he was divinely appointed. This might be part of the reason that Genghis Khan, after subjugating a people, would do things to make life better for them.

Though many considered Genghis Khan's military campaigns as a great slaughter and destruction at an unprecedented scale, others considered him a liberator. The estimates of the number who died in his conquests is a huge range from four million to sixty million [18].

However, after people were conquered, Genghis Khan lowered taxes, created a postal system, distributed wealth, established just laws, and granted religious freedom. He also enabled ways to share the best each group of conquered people had invented, created, or written with others in his kingdom [19]. These practices, in turn, made the people love him more and helped them feel he was even more divinely installed into the position that he held.

All of these, taken together, the king being divinely appointed, the source of all good from heaven, and the unifier and benefactor of society, instilled a belief in the people that he should be protected. This, then, became one facet of the reasons why the army would fight to the death to protect him.

Loyalty and Forced Protection of the King

One final concept for why the army would fight to the death deals with loyalty and forced protection of the king. In the exhibit about Genghis Khan, there was much about how he developed his followers and his army. Their society followed the concept that if someone has something and you can take it, it was okay to do so. This came with the same belief that God had given you the power to take it from your enemies; therefore, He was okay with it.

The leaders would collect the spoils of war, and then use them to buy loyalty. Nibley says that leaders got associates by bribery and kept them with oaths. Often the men would flock to the banner of the winning leader, partly because of the draw of wealth expected from the spoils of war. This purchased loyalty was not always the strongest, and such could change quickly [5]. A continual stream of wealth was needed to keep those connected to the leader in this way from turning to someone else. This can be seen in the Book of Mormon when the sons of Akish, bound to their father by oaths and lavish gifts, used the same means to create followers of their own and then started a war with their father (Ether 9:11-12).

That is one of the reasons why someone like Genghis Khan had to continue his conquest of more lands. The number of items they would have to give as gifts would eventually dwindle and need to be replenished. Therefore, there would need to be a conquest of more people in order to refill the king's treasury [19].

However, there was one group protecting the king whose loyalty was even less than those from whom it was purchased. This would be the group that was forced to fight. In the Genghis Khan exhibit, those captured in battle, especially those who unwillingly surrendered, were forced to fight in Genghis Khan's army. These people often retained animosity at their subjugation. Giving people who felt this way toward you a weapon could be dangerous. They could just as easily turn it on you.

For this reason, most were put at the front of the battle and were forced to bear the battle rams to break down walls, or use other such devices that could not easily be turned against the king and those loyal to him. Being at the front line of the battle, they were usually the first ones killed [19].

The king would surround himself in layers with those who were most loyal closest to him, and with each outward layer being less and less loyal. This culminated in those who were least loyal being farthest away at the front of the battle. Before the enemy could reach the king, they would have to destroy those who were least loyal, then fight their way through his more loyal inner circles. Understanding this method of army placement helps us see how an entire army can be obliterated to save the leader. The least loyal are destroyed first, while those who are most loyal, and willing to fight to the death to protect the king, are the last to die.

This concept of warfare can be seen in the game of chess. The pawns are put at the front of the battle. The word pawn is defined as "a person who does not have any real power but is used by others to achieve something" [20]. Those closer to the king on the chessboard, such as the queen, the knights, the bishops, or the rooks, have much more power and are used strategically to protect the king.

Why do they prefer to capture the opposing king, and what do they do with him?

This is one of the strangest parts of the book of Ether. The appendix in Table 3 - King Succession shows that it is almost equal for the opposing leader being killed as being put into captivity, with captivity having the slightest edge. In our society, we would kill or capture the enemy. So those two things are not what seems strange. What is unusual to us is what this captivity is like.

For us, if someone were captured, they would be imprisoned and given no preferential treatment. Perhaps they would even be interrogated or subjected to less than favorable circumstances.

But in Ether, we read that the captured leader has children in captivity. At times there are enough children born in captivity that it appears the captured leader probably has multiple wives. According to Nibley [5] having multiple wives was a sign of wealth and power. This would mean the captive leader would have some form of wealth and power even in captivity.

Even if this is not the case, it is strange to us that their life was such that they had children in captivity. This shows that they were not just locked in some prison, or at least not always locked in a prison.

Further, in Ether 7:7 we read:

And it came to pass that Kib dwelt in captivity, and his people under Corihor his son, until he became exceedingly old…

Depending on how you read this, Kib's people were also in captivity, or at least under the rule of Corihor, his son. But it appears the people were in captivity, and it is not reasonable to think that they were all in prison. This captivity was probably more of a servitude where the people had to pay taxes, perform work, or provide goods to Corihor.

The enslavement of the people of Israel in Egypt, and those from Judah taken to Babylon, were both called captivity. In both cases, they were considered servants of the king and were forced to work at his bidding. The level and brutality of this servitude could have ranged from workers in his fields, on building projects, or in his palace, to enslaved people working at the end of a whip. But these people, in their bondage, also had children.

The one difference that makes it unusual to us is that we probably would have locked up the leader. But apparently, he was forced to serve like anyone else, except that he might have had more wives, wealth, and honor. The book of Ether, at least once, uses the phrase "serve in captivity" (see Ether 10:15).

Sometimes, as in Ether 8, there is no mention even of captivity. Jared goes to war with his father, Omer. Omer served in captivity for half his days, but he had sons and daughters while in this captivity. Two of these sons became angry with their brother and raised an army. After they had destroyed his army, Jared begged for his life, saying he would give up the kingdom to his father.

It makes you wonder how he could not give up the kingdom with no army left. However, there might be something to this. Even though the army is gone, there are still the ordinary people to consider. Perhaps if the king is allowed to live, he must acknowledge the new king so his people will do the same. Referring back to the movie Mulan, remember how Shan Yu told the emperor to bow to him. He was telling him that he must acknowledge Shan Yu as the new emperor so the people would follow suit. When the emperor refused, Shan Yu was going to kill him.

Continuing the story with Jared, once his brothers allowed him to live, there was nothing more said about captivity or anything for him. In fact, Jared, at his daughter's urging, is soon having a feast and inviting Akish to join in to plot the overthrow of his father again. (You might note here that it says Akish is a friend to Omer, Jared's father, not a friend to Jared.) But the main point here is that after all that Jared had done against his father, there was no death or prison or anything to stop him from doing this again.

It is also important to see that Akish, a friend to Omer, was quick to abandon that friendship for a lustful desire. It could be that Akish was the type of friend bound by gifts, greed, and oaths, as has been previously mentioned. Men connected to the king in this way, offered the same things from someone else, tend to quickly dissolve the previous friendship [5].

But we still haven't discussed why the person is allowed to live, or why they have the type of captivity where they can have children.

Nor have we discussed what that captivity might be like and why. From my research, there appear to be many reasons why a person is put into this type of captivity instead of being killed.

The Captured King is a Relative

In evaluating the relationships of those put into captivity in Table 3 - King Succession in the appendix, something interesting appears. Those who are named that bring the king into captivity instead of killing him are related, except in one case. The ones that are not named that do this could be related, but we can't tell from the record.

In the one case where a named person, not known to be a relative, brings someone into captivity, it is Shared bringing Coriantumr into captivity. Nibley [5] says that the two are "royal brothers" (p224). I cannot find anything that says that. Perhaps Nibley assumes it since Shared does not kill Coriantumr. Later, the sons of Coriantumr fight Shared and beat him but don't slay him. He could have fled, but nothing is mentioned. Coriantumr regains his throne, and later he again fights Shared. The battle wins go back and forth until Coriantumr slays Shared.

But one thing that makes me think Shared and Coriantumr were not brothers is that the person who rose to fight against Coriantumr next was Gilead, whom Ether says was the brother of Shared. If Coriantumr and Shared were brothers, it would be likely that Coriantumr and Gilead would be brothers too, but the record says nothing about it. (Of course, with multiple wives, deaths, and remarriage, it is possible that even with Coriantumr and Shared being brothers, Coriantumr and Gilead were not.)

Nonetheless, in every other case where the person is named and puts the defeated king into captivity, the two are related. Sometimes even when the person isn't named, the record indicates they are related when they put the king into captivity.

There are times when relatives put the person to death. But I find it interesting that the closer they are related, the more willing they seem to be to put them into captivity instead of killing them.

In the exhibit on Genghis Khan, a section talked about how he dwelt with those he defeated. One key note was that in the cases of those previously captured revolting against him, he often destroyed them. One prominent exception was the brother of one of his favorite wives. He killed everyone else involved in the rebellion but let her brother live.

This makes a strong case for one reason the defeated king is not killed is because the conqueror and the conquered are related. It still appears there are other factors, but this is likely an important one.

King Divinely Installed

As was discussed previously, the people felt that the king was divinely installed into his role. To kill someone of royal blood could be seen as a great offense against God, even if they were the enemy.

A good example is Genghis Khan and his best friend, Jamukha, as shared by Weatherford [19]. A little side note about Genghis Khan and the Mongolians is of value here.

Until recently, little was known about the Mongols. A big reason for that was there were few written records, and those that did exist, the Communist leaders of Russia tried to suppress. There seemed to be a fear that understanding more about them would spark a nationalistic movement among the Mongolians.

However, there was a written record in China, written in Chinese. It had become known as *The Secret History of the Mongols*. This was a record that was apparently created in Genghis Khan's court. But even though it was written in Chinese, it was only the sounds of Chinese, and those sounds meant different words in Mongolian, so the translation made no sense. Russia kept foreigners out of Mongolia, which made it impossible to learn what those sounds meant in the Mongolian language.

But with the opening up of Russia in 1991, suddenly there was more ability to understand the meaning of this record. As scholars started translating the work into other languages, a whole new understanding came about the Mongols. For centuries, people had viewed them as nothing but barbaric killers, but that was not the case. Though Genghis Khan and his descendants killed many people in their conquests, those who were left were given religious freedom and a society with the best of what was found in different parts of the kingdom. This even included things like a postal system. Weatherford even attributes much of the Renaissance to the innovations coming out of the Mongol-controlled world [19].

Part of this record included the story of Genghis Khan. The record was not really written for those outside his family, but it gives an intimate look into his life.

Genghis Khan's given name was Temujin, and his best friend was Jamukha. Jamukha and Temujin even swore a public oath to each other as blood brothers. They started off their conquests together. However, even though they were distant cousins, Temujin's mother was a captured woman from another tribe. After his father died, she was forced out of her husband's tribe. She was a brave, strong woman who tenaciously worked to provide for her children [19].

Figure 25: Genghis Khan and best friend, Jamukha, with Toghrul Khan's army

Because of her status, Temujin (Genghis) was of lower birth, and after a while, Jamukha took over total control of their army's leadership and started treating Temujin in a subservient way. Temujin's wife suggested he leave and start his own tribe, so he did.

This kind of action can be seen in many parts of the book of Ether where someone who is dissatisfied with the status quo goes off into the wilderness and creates his own group. After his group has grown sufficiently, he goes to battle against those he left.

Tamujin's group grew to the point he took on the title of Genghis Khan. He eventually did come to battle with Jamukha and his followers, and Genghis lost and had to retreat. But Jamukha was terrible to those he captured, so much so that some of his own followers turned away from him. Also, Jamukha ruled tribes he conquered by nobility of birth, while Genghis ran his based on meritocracy, meaning the better a person worked and performed for the people, the higher he would rise in office. This led to a split, with more common people joining Genghis and more royal people following Jamukha.

Eventually, Genghis beat Jamukha, but instead of killing him, requested he join in Genghis's army. But Jamukha, apparently somewhat ashamed of how he had treated his friend, requested as a final honor that he be killed. He said he would support his friend from the other side. He asked that his bones be placed on a high mountain that he might watch over Genghis [19].

This was hard for Genghis, but he did as requested. One thing that is strongly stated is that ". . . they kill him in the aristocrat way without shedding his blood on the earth without exposing it to the sun and sky" [19](p.64). It was considered a bad thing to spill royal blood if it could be helped. Again, this went back to the idea that God had put the person in their position.

The Two Crownings and Three Respects

This final concept of why a captured king was allowed to live can be found in a concept in ancient China called erwang sanke or "The Two Crownings and Three Respects" [21]. It has been challenging to find information on this tradition, though there has been brief mention of it in multiple documents about Chinese history.

A typical Chinese practice anciently was to grant land and nobility status to members of overthrown dynasties [22]. This was known to occur in many dynasties. The conquered nobility were given lands to govern, kept royal titles, and were given the privileges of royalty, which would include multiple wives and children. One key ingredient in this was that they would rule the land they were given, subject to the overall king.

The phrase "The Two Crownings and the Three Respects" is assumed to come from the past and present dynasties (the two crowns) and the fact they would honor three generations of the past dynasty. That is, the children, grandchildren, and great-grandchildren (the three respects) would receive the honor of these nobility titles [21].

In the readings, there are questions about the reasons for this practice. Multiple reasons were given. First, it made the changeover to the new ruler less dramatic and more acceptable to the people. Second, it split out the governing process and made it so the previous rulers were expected to keep their own people in line. Third, it appears to be related to the concept given previously that God installed the rulers. The final reason is especially only hinted at, and it is much harder to find a full explanation of it. It appears that keeping the past rulers in power, with the new king ruling over them, gave the new king added prestige and legitimacy as being installed by God.

A statement related to The Two Crownings and Three Respects is attributed to Khubilai Khan, a descendant of Genghis Khan. When he conquered a nation, he sometimes left the ruling party in power to rule subject to himself. He let them keep the affluent life they had been used to. When asked why, he said that if people were allowed to maintain the life they were used to, they would have less of a tendency to rebel. However, if their living standards were reduced, they would not be satisfied with their state in life and would foment dissent [19].

Let's look briefly at each of these ideas.

Lessen the Impact of Changeover to New Rule

As a new dynasty came into power, often there were many conquered peoples to deal with. Bringing them under one head with laws and governing that differed from what they were used to could invite rebellion. One thing the people seemed to appreciate was consistency. Having the previous ruler continue to rule as he had, helped the people feel much less change than they would if laws and rules they weren't used to were immediately implemented. By letting the previous ruler rule, life could go on much as before.

Split Out the Governing

As the kingdom grew, trying to rule the many different parts and different people became a monumental task. Allowing a ruler to govern those he was used to governing reduced the burden on the king in managing the affairs of all areas. The previous ruler did have to swear loyalty to the new king as well as pay taxes, tribute, or whatever else was required. But he was the one who had to answer to the new ruler if something went wrong or if the promises were not kept.

Rulers Installed by God

This part refers to what was talked about previously; if the new ruler was to be considered as being put in his position by God, he would have to assume the previous ones were as well. By treating them with respect as nobility, he could claim the same for himself. This idea of The Two Crowns and the Three Respects then became a perpetuating concept from dynasty to dynasty until the idea was abandoned in favor of putting people in positions of authority based on their loyalty to the king. Genghis Khan was one of the foremost to make this transition [19].

Another part of this was that the past emperors were considered to have become god-like when they died. By letting their descendants still honor them, they could bring desired blessings to the people and the new emperor from the heavens.

Added Prestige to the New Ruler

This last concept is harder to nail down and explain. A big reason is that I feel it is more implied than stated in the books and articles I've read. I don't have as much backing for the idea from other sources. Therefore, I take responsibility for it and suggest the reader consider it and evaluate its merit.

To explain the concept, I would like to use an analogy. Suppose you are in a high school and want to show that you are the toughest person there. If you want to show everyone that you are indeed the toughest person, you can't do that by picking on the weakest, smallest person in the school. You would have to show it by defeating the person everyone else felt was the toughest.

It reminds me of when I was a team captain on my high school wrestling team during my senior year. Nothing upset me more than when one of our younger wrestlers mocked and pointed a finger at a wrestler they had just defeated. I would always pull the younger wrestler aside and let them know that mocking someone, and belittling them, only made the person doing it look smaller. If they showed respect to their opponent as someone of exceptional ability, then their win was much more respected. I would always tell the young wrestler that I never wanted to see them disrespect their opponent again.

For the conquering kings, there is the idea that if an enemy is considered extraordinary and powerful, and you can beat them, then you are considered greater still. But if the enemy is considered weak, and you beat them, it adds little to your prestige.

This is a possible reason that a beaten king would be allowed to keep a royal title along with royal responsibilities, honors, and pleasures. They were allowed to continue to be great royalty, but they must bow to their new ruler. This concept can be seen in the Disney® movie of Mulan, where Shan Yu tells the emperor to bow to him. This would also have the added effect of expecting the people the emperor ruled to do likewise.

Therefore, part of the reason for The Two Crownings and the Three Respects is that honoring the beaten emperor makes the current emperor seem greater.

Why do the people accept the new king?

There are a few reasons the new king was immediately accepted. If all of the army except for the king was destroyed, then there was no fighting force left to defend the people. It would then be understandable for the people to accept the new king's authority. On the other hand, when the king was killed or captured by a commando-type squad, few, if any, in the army were destroyed. This would be when things seemed unusual to us.

Imagine if some enemy sent a commando squad into the Whitehouse here in the United States and killed or captured our president. If the leader of that group declared himself to be our new leader, we wouldn't put up with that at all, but would send our military after them. However, our government differs from that of the ancient Orient or of the Jaredites. The power in our government is not vested in one man.

When power is invested in one man, or even one small group, it is much easier to destroy that governing power and install something or someone else. Still, it seems strange to us that in such a case, the new king would be accepted without question.

However, when you consider the belief of a king's divine right to rule, and assume he was installed in the office by God, it becomes easier to see why they would accept him. If they believed God appointed the previous king, then it makes more sense to assume that if someone overthrew the king, even with a commando squad, then that person was divinely appointed to take his place.

Nibley [5] points out that one challenge for a king was that if he could do this to the king before him, he had to always be careful because someone else could do it to him. A good example of this is in Ether 8 and 9 when Akish slew his father-in-law, Jared, and put himself on the throne. Jared, himself, had come into power the second time by plotting the same thing for his father, Omer. In fact, as has previously been pointed out, Akish was Omer's friend, not Jared's (See Ether 8:11). Yet Akish was quick to join Jared for the promise of lust and wealth.

Akish seemed to be aware of the vulnerable position that he was in as king. It apparently made him suspicious of anyone who might take power from him. In Ether 9:7 it says:

Ether 9:7 - And it came to pass that Akish began to be jealous of his son, therefore he shut him up in prison, and kept him upon little or no food until he had suffered death.

We are not told of the reason for this jealousy, but it is likely because his son was becoming popular with the people or had some other means whereby he might overthrow his father. This is a valid fear since others of his sons formed an army to fight against him.

One challenge Genghis Khan had with uniting the people was that they would change loyalties in an instant, usually toward whomever was winning. The people wanted to be on the winning side. This is understandable from a safety standpoint, but it was also for the ability to gain reward, including looting others. We might not call this loyalty at all, and we still see this today in some cultures.

A friend of mine who served in Afghanistan said that one of the biggest challenges of being there was their cultural norm of switching loyalty, as he called it. He said they would free a village from the Taliban, and the people would act grateful. The village leaders often swore their loyalty and help to those who had just cleared the Taliban out of the town.

However, my friend said that if the Taliban came back and seemed to have the upper hand, the people, including the village leaders who had promised their help, would turn against those who had previously driven out the Taliban. This is different for us as we view loyalty as something that transcends the moment, even if it means death.

We would call anyone who switched their loyalty like that fickle, disloyal, or a traitor.

My friend said that the problem this caused in Afghanistan was that they never knew who their enemy was. Someone who was a friend and ally one day might be your enemy the next. They could fight by your side or give you info about the enemy one time, only to be the ones shooting at you soon after. He said a person had to assume every Afghan was the enemy every time.

This changing loyalty was the thing Genghis Khan hated the most. That was why he was more likely to put someone to death for it quicker than for any other reason. If someone defected from the other side to his army, he didn't trust them. He knew that if they were disloyal to his enemy, they would be disloyal to him. Even though he might integrate them into his army, they often would be in the most vulnerable positions, far from him [19].

So, this cultural norm of switching loyalty to the new king, though strange to us, was and still is a strong tradition in some of those countries. Understanding that, and coupling it with the belief in a king's divine right to rule, it is easy to see why the people would immediately change their loyalty to the new leader, no matter how he came to power.

7

Other Unusual Concepts of the Book of Ether

＋

In this chapter, I'd like to address a few interesting items in the Book of Ether. Some concepts in the book seem unusual for the time, or are not mentioned when we would think they should be. I have also had questions regarding some of these asked in my Education Week classes. These items include:

1) The mention of steel and windows of glass
2) Silk is mentioned in Ether 9:17, yet it doesn't appear to be in the Americas when Columbus came.
3) Animals not found in the Americas when Columbus came, including elephants, asses, sheep (there might have been some types like mountain sheep), cows, oxen (depending on meaning), horses, and swine (Ether 9:17)
4) Animals we are not sure about, including cureloms and cumoms
5) Honeybees mentioned before crossing the ocean but not after. (Honeybees came after Columbus.)
6) Passing of who would be the next king is not always to the oldest son

Steel and Windows of Glass

Nibley [5] says that scholars often mocked the idea in Ether 2:23 that the Jaredites could have known about windows that could be dashed to pieces. They claimed glass wasn't invented until much later. But the problem is, glass is made of sand and deteriorates quickly, so if ancient civilizations made glass, there would be little or no evidence of it.

We now have proof that glass-making was known over 4,000 years ago [23]. That would put it near the time of the Jaredites. However, because of its deterioration properties, it could have been known earlier than that, and we don't have evidence of it. This is a case where science seems to be catching up to what the scriptures say, and a lack of evidence does not constitute proof.

Just like glass, some people mocked the idea that the Jaredites could know about steel (Ether 7:9). However, steel has now been found dating back to around 1800 BC, not long after the time of the Jaredites [24]. So, it is very plausible that it could have been known earlier, and we just haven't found it. Steel was hard to make, especially anciently, and people would take good care of the items they had. No one would just lay something around so it could be discovered in the future.

Silk

Silk is mentioned in Ether 9:17, yet it doesn't appear to be in the Americas when Columbus came. However, there are claims silk was known in China from 4000 BC [25]. This means it would have been known at the time of the Jaredites. Something like silk, that people greatly desired, and was easy to bring to America, isn't hard to imagine having come with the Jaredites. So, why was it lost in America and didn't exist at the time of Columbus?

Figure 26: Silk Worm

There could be a few reasons for this. In China, the production of silk was protected. It was illegal to take silkworms out of the country at the peril of death. However, the history of silk claims that a princess that was promised to a prince of Khotan, a land farther west, didn't want to live without silk and smuggled some silkworms out of the country.

Perhaps those who raised silk here in the Americas also tried to keep control of it. But why would the silkworms have disappeared?

A couple of possibilities could be drought and destruction. It is possible that the silkworms could have been lost during one of the great droughts, though silk production is mentioned in Ether 9:17, right after one great drought. Still, weather or other natural phenomena could have been a problem. This includes diseases which have threatened the industry at times [26].

Also, with the war that finally destroyed the Jaredite culture, it is possible that the whole silk industry was destroyed by the burning of cities and villages. Man is often the culprit in the destruction and extinction of elements of nature. An example was the nearly complete annihilation in the 1800s of the once plentiful buffalo in North America. But the point is, just because the Book of Ether mentions something that was not found when Columbus came, it is not sensible to suggest it could not have disappeared by then, since we have similar things that have gone extinct locally or globally in our day.

Animals Mentioned but Not Found When Columbus Came

The animals not found in the Americas when Columbus came include elephants, asses, sheep, cows, oxen, and swine (Ether 9:18-19). Tradition says the horse wasn't here and was brought by the Spanish However, there is evidence that horses existed, at least among the North American natives at the time of Columbus's arrival---just not the Spanish horse [27]. There were animals like Bighorn Sheep that were here at the time of Columbus, but this scripture seems to talk about domesticated animals, and Bighorn Sheep would not be easy to domesticate.

Animals like cows and oxen can be interpreted as different types of the bovine family. But it is still unlikely that the animals described in this scripture were any of the types found in the Americas. Where did these animals go?

First, it must be noted that some of these animals appear in the Nephite record. The first time horses are mentioned among the Nephites is in Enos 1:21. That is also the first mention in their record for cattle and goats. There is also the use of the phrase "flocks of herds" separate from cattle and goats, which probably is a reference to sheep. However, the existence of asses, elephants, and swine is never mentioned outside Ether 9:18-19. That does not mean they did not exist among the Nephites.

Just because something isn't mentioned does not tell us they did not have them. When things become common or mundane, they could have been left out of the record. But I would think at least something as significant as elephants would be recorded if they did have them.

Just as with the silkworms, these three could have been lost through destruction or used as food during famines or wars. This destruction may have come during the time of the Jaredites or later with the Nephites. Those mentioned in the Nephite record could have also been lost from them for the same reasons after the Jaredite nation no longer existed.

It is important, as with the silkworms, to understand that just because something was not found at the time of Columbus, that doesn't mean they weren't here before he came. More and more science is showing that there were creatures here that aren't now. Though science puts some of them as more ancient than the Jaredites, more evidence is coming out all the time of their existence. This includes not only the elephant, but the camel and other things [28].

Figure 27: Animals mentioned but assumed gone before Columbus, or possibly mentioned but identity not known in Ether

Cureloms and Cumoms

There are a couple of animals mentioned in the book of Ether whose identity we don't know. These are the cureloms and cumoms (Ether 9:19). We have no such animal names to identify them with in our day. The only additional information we have is that the Book of Mormon names them along with the elephants as the most useful animals. As far as I know, Joseph Smith did not elaborate any more on them, which likely means it was not revealed to him. What could they be if they still exist, or what could have happened to them if they don't exist?

It is possible that they could be extinct species which we know nothing about. They could also be animals we call by different names. If they are something we know, or know about, what could they be?

The fact that they are grouped with the elephant might mean they were a beast used for burdens and work, which is one of the main things elephants were used for. An animal that matches that in the Americas might be the llama or alpaca. When I was in Peru for part of a summer with the faculty of Brigham Young University – Idaho, the llama, and its relative, the alpaca, were used for just about everything. As we traveled by train to Mauchi Pichu, there were often people leading llamas laden with loads of sticks, food, and just about every other imaginable thing they could carry.

These trains traveled so slow, it felt like we would never get to our destinations. One of the train tour guides told us there was a story of an old man with a bundle of sticks on his back walking alongside the railroad tracks. The train conductor felt sorry for him and asked the old man if he would like a lift.

"I would," the old man said, "but I'm in a hurry."

While we traveled, the people who ran the trains hired college age students to try to sell us stuff. This included modeling Peruvian clothing, jewelry, and other such items. Maybe that is why the train went so slow. It gave them more time to talk us into buying things we didn't need.

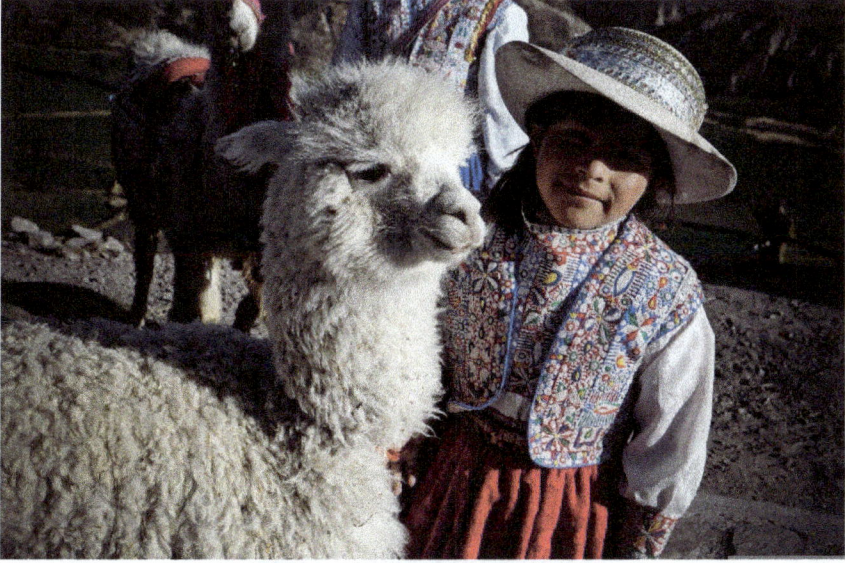

Figure 28: - Peruvian girl

Also, the main dishes on every menu in the restaurants we ate in was llama or alpaca meat. (Guinea pig was also there but was considered a delicacy and was very expensive.) There were a couple of fun stories with food there that might be of interest here.

We had been in Peru for some time, and I was getting a little tired of llama and alpaca in everything. One day I was in a restaurant and saw the word "tuna" on the menu. It was lunchtime, and I wasn't feeling like eating anything heavy, so I asked the waiter if the tuna could be made into a sandwich. He looked at me in surprise and said, "If you like. Anything for you, Señor." He then asked how I would like it.

I told him I would like it with plenty of mayonnaise between two slices of bread. Again, the shocked look on his face made me realize that was unusual there, but he said he would do it. My colleagues all ordered the usual llama and alpaca meat meals.

When our food came, I realized the problem with the language barrier. Tuna in Peru is the fruit of the cactus. It tastes somewhat like a semi-ripe watermelon. My sandwich consisted of lots of bread, lots of mayonnaise, and a large slab of fruit. It didn't look appetizing at all.

Figure 29: Cactus (Tuna)

My colleagues, realizing my dilemma, had a good laugh. It didn't taste any better than it looked, but I told them it was wonderful to reduce the good-natured teasing. And I did eat the whole thing, not because I liked it, but because I was hungry and was too cheap to buy something else.

Later, at the end of our time in Peru, we ate at the same restaurant, and I noted that penciled in at the bottom of the menus was a new item: "Tuna Sandwich." The description of it was two pieces of bread with a good helping of mayonnaise and tuna inside. We got the same waiter as before, and he was pleased to tell me that my "creation" was very popular with foreign tourists.

"But," he said, "they not eat much. Must be thing to order when not too hungry."

I had told this story to my wife, Donna, and one day we were in California and stopped at a little roadside fruit stand. There was a Mexican boy and his father there. Donna saw a sign that said, "Cactus Fruit," and remembered my story. She called me over. The father didn't speak English, so she asked the young man what it was, and he said, "cactus fruit."

"But what is it called in your language?" she asked.

"Oh," the boy said. "We call it tuna."

Then she got a chance to experience it firsthand. It is actually pretty good, but not with mayonnaise.

A second fun story about food in Peru occurred just before we left to come home. We went to a restaurant, and there were all the typical alpaca meat dishes. But one item in big letters stood out from the rest: "American Hamburger and Side Salad." I decided to order it.

"Are you sure that's a good idea after your fiasco with tuna?" one of my colleagues asked.

"What can go wrong with an American hamburger?" I asked.

In Peru, bread is a pre-meal thing. They would always bring us lots of bread to eat while we were ordering and waiting for our food. But the minute our food was ready, the bread was cleared away. I was even told it was considered improper to have bread with the meal.

While we waited for our meals, we ate bread and talked about what we had eaten while in Peru. Finally, our meals were ready, and true to form, the waiter cleared away the bread. When my meal was placed before me, there was one hamburger patty, and beside it was a stack that consisted of one tomato, one onion, and a piece of lettuce.

"I hope you enjoy," the waiter said. "Our cook just come back from United States. He learn all about American food."

He left and all of us at our table considered my "hamburger."

"Apparently, the cook must think that the tomato, onion, and lettuce for your hamburger was a side salad," one of my colleagues said.

"When I lived in New York, they often served it that way," I replied. "That way you could put on the things you wanted. But my question is, where is the bun?"

I called the waiter over and asked him that question.

He looked confused. "I not know American food. I ask cook."

A few minutes later, he came back and said, "Cook say decorative roll served with hamburger in United States cost extra."

When he left, my colleagues laughed. "Obviously, because they don't serve bread with the main meal, the cook thought the hamburger bun was just a decoration," a colleague said.

I then realized that my request for a "tuna sandwich" was probably even stranger to them than I realized, because I had wanted bread with the meal. I was nearly out of money for our trip, so I just ate my "American hamburger" without the bun instead of paying extra for it. I wrapped my side salad around the meat and ate it that way.

But what has this story with a side salad of fun have to do with the Jaredites and cureloms and cuemoms? Some people have suggested that maybe these animals were things like the jaguar, monkey, or other things. But where they were listed with the elephant as the most useful items and in the same verse as horses and asses, I would think they likely were beasts of burden, at least one of them. The llama, alpaca, and other related animals are used as beasts of burden, but they are also used as a primary meat source—at least in Peru.

As I worked further on this, I searched Sumerian references for anything related to these words or parts of these words as possible prefixes and suffixes. Sumer was an area in Mesopotamia where possible ties could be found to the Jaredites. Their culture predated the Babylonians. However, no one knew they existed until relatively recently [29].

There have been tons of clay tablets found in the Mesopotamian valley. (Literally tons. They are very heavy.) Many of these had writings that could not be interpreted. In the 1800s, a pillar was found with different writings. Many scholars thought they were different things. But, as multiple people worked on the translation, they eventually realized the various writings were of the same thing, like the Rosetta Stone. Working on these ultimately made it possible to translate them and gain the understanding to translate the clay tablets and other writings. These translations brought to light the culture and people of Sumer who were not known before that time. The abundance of tablets and other writings gave us more understanding of them and surrounding areas, like Akkad and Elam, than we have of many ancient civilizations [29].

Sumer and Akkad were the area of or nearby the Tower of Babel and included the city-state of Kish. These translations gave us more insight into the culture the Jaredites came from than we would have had otherwise. Figure 30 shows a map of the area where these civilizations were.

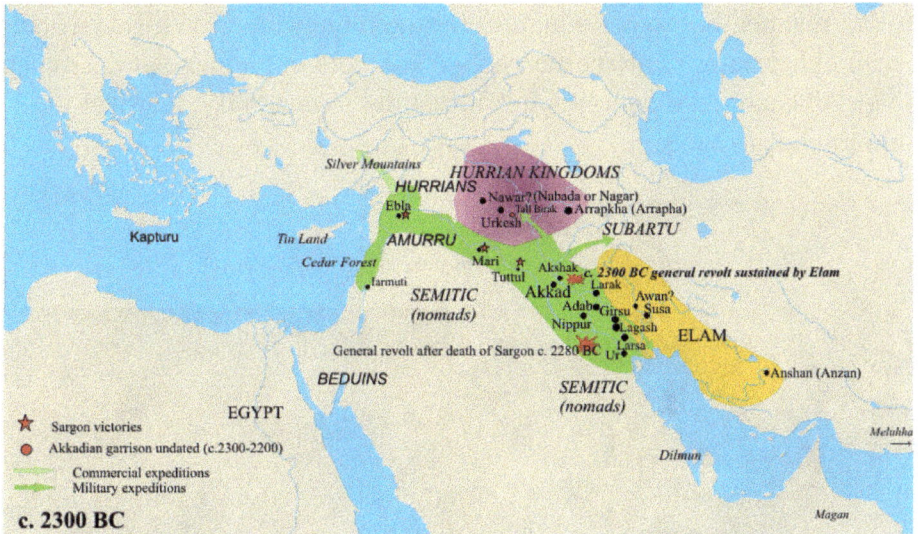

Figure 30: Area of Sumer, Akkad, Elam and other ancient civilizations at the time of the Jaredites.

The study of Sumer led to an article by Jerry Grover [30].

Grover talks about how the combination of words in Sumerian can give meaning to items. Kur or kura means mountain (as in the cur in cureloms). He also says the word *e3-li-um* (a form of the word *e* meaning sheep) with kura (based on sheep) means looms. This could refer then to "mountain sheep" but probably not the bighorn sheep that we often call that. It would be more like a mountainous animal that was fur-bearing. This fits well with the alpaca. (I even brought back a nice sweater made from alpaca wool from Peru.)

Regarding the cuemoms, Grover discusses the words ku (plow) and *u2-ma-am* (form of word *umamu* meaning beasts). So cuemoms could mean plow beast. This might fit the llamas which are used for all sorts of burdens; at least they were in Peru.

I will relate one fun story about alpacas. When I was in Peru, the Peruvians liked to dress in colorful native costumes with an alpaca by their side. They would charge you money to take their picture. One lady had herself, her alpaca, and her cute little daughter, who was about four. I asked if I could take their picture and she said, "One sol or one dolla." That was about the only English some of them knew.

And as I mentioned before, I could get three sol for a dollar, so I always kept a pocket full. I pulled the sol out of my pocket, and as I reached it out to them, the little girl took it. After I took the picture, the

81

mother reached out to her daughter for the money, and the little girl held it behind her back. The mother sighed and held out a piece of candy, and the girl exchanged the money for the candy. Apparently, that was a common occurrence.

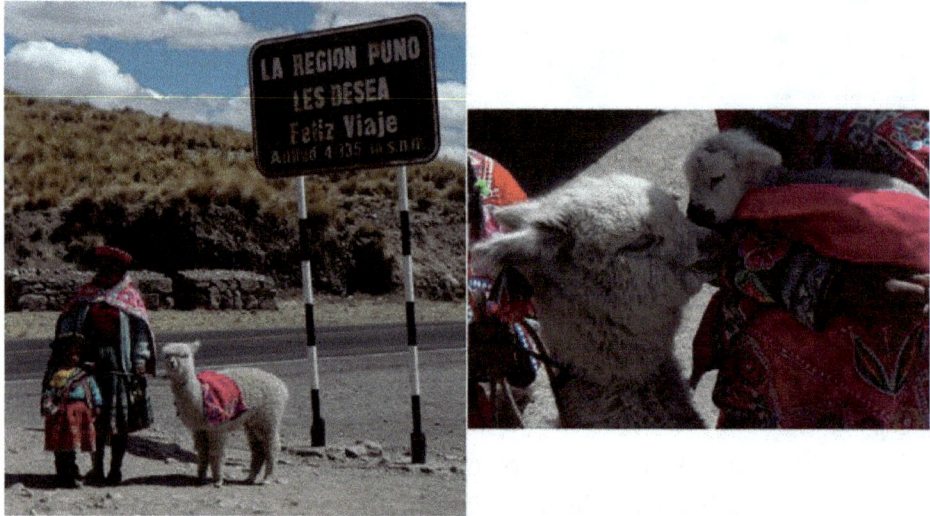

Figure 31: Personal Pictures from Peru Trip

I had some other thoughts on this. These are some of the most useful animals in South America, but probably the leading animal for usefulness in North America to the native people was the American Buffalo. The tribes of the planes relied almost solely on it for a source of food. The only question on that would be that the Book of Ether seemed to be talking about animals that were domesticated. The buffalo is a hard animal to domesticate, though I have seen one that was. I have a friend who had a buffalo that he raised from a calf. He used to ride it. But it went where it wanted, and he was just along on top of it for fun. There wasn't much he could do to dictate its direction if it didn't want to go that way.

I doubt we will ever know for sure in this life what the cureloms and cuemoms are, but there was one other thought that I had. Their names are close enough to each other, that they might be related or similar animals. This gives more credence to the fact they might be something like llamas and alpacas that are related. It will be interesting to ask Ether when we get to the other side.

Honeybees

The honeybee is another interesting item. Ether 2:3 says:

And they did also carry with them deseret, which, by interpretation, is a honeybee; and thus they did carry with them swarms of bees, and all manner of that which was upon the face of the land, seeds of every kind.

Figure 32: Honey Bee

But no honeybee is known to have existed in America at the time of Columbus. The colonists brought them when they came in 1622 [31]. That doesn't mean there weren't bees here. There are about 20,000 bee species, and about 4,000 are native to the Americas. There is one bee that is a tropical bee that produces honey that is native to America, so there was honey in the Americas, just not at the rate honeybee produces.

So, what about the honeybees the Jaredites had? Nibley [5] claims that the honeybee was considered sacred, even to the point of worship in ancient cultures. It is hard to believe if they worked to take them on their journey that they wouldn't have brought them on across the ocean. But in the Jaredite record, honeybees are never mentioned after the Jaredites build their boats that are "tight like a dish" and set off for the promised land. The big question is, did they ever reach the American shore?

If any bee colony had made it to the Americas, it is likely that the spread of swarming bee colonies would soon have been all over, as happened when the early colonists brought them in 1622. So, what happened to the Jaredite honeybees?

The first possibility could be that they never brought them into the boats. Honeybees do sting, and putting them in an enclosed boat could present a possible danger. A person can visualize the Jaredites saying, "You put them in your boat." "No, you put them in yours."

There is always the possibility that with eight vessels, everyone thought someone else had taken them on their boat, and they were forgotten. In that case you can imagine them getting to America and saying, "All right, who forgot the honeybees?" "Well, don't look at me; I'm not going back for them."

Besides the fact that they may never have gotten on the boats, there is the possibility that they didn't last through the voyage even if they were on the boats. Honeybees forage to survive, but in the winter and times of limited food, they eat the honey stores. It is possible that in almost a year on the ship, the bees ran out of food. In comparison, the voyage time for those crossing the Atlantic in 1622 when the colonists brought the bees to America was usually just over two months. This would make it easier for the bees to survive the voyage. Plus, they had open ships.

One challenge with this idea is that the Jaredites brought enough food for their own survival. That should mean they would have had some idea of how long it would be. If that was the case, if they brought the bees, they should have taken enough food for their survival. But humans can eat fish, and there is the possibility they supplemented their food supply that way. Bees can't eat fish.

It is also possible that they brought the bees and somehow lost an important part, like the queen. Bees can't regenerate without the queen. There are a myriad of reasons that they may have had that the bees didn't make it here.

There is another possibility. In modern times, we have seen enormous problems with bee diseases, such as the colony collapse disorder. These diseases are especially acute in the honeybees because of their social structure [32]. The bees native to America are more solitary. It is hard to imagine a disease that could wipe out all the honeybees on the continent, but it could have happened that the bees made it here, then disease wiped them out.

All we know is that the Jaredites don't mention the bees after leaving for America and that they weren't here when Columbus came. But they did exist in China, so maybe the Jaredites left them there.

King Succession Not Always to Oldest Son

In many, if not most, cultures that are autocracies, the succession of the kingdom is from the father to the oldest son. This is not always the case in the record of the Jaredites. A different son was often chosen. Of course, there are also times when a son, ambitious for power, takes matters into his own hands and dethrones his father, taking ruling for himself. We have seen this in the example of Jared, the son of Omer. But when the king chooses his successor, what could be the reasons for choosing someone other than the oldest son?

Ether 9:14 And it came to pass that Omer began to be old; nevertheless, in his old age he begat Emer; and he anointed Emer to be king to reign in his stead.

Omer had sons and daughters previously who fled with him. Why is this younger son made king? Perhaps, with all the problems they had seen their father go through with Jared trying to destroy him, the others did not want to be king. Being a king could be a grueling task.

In Ether 7, Shule, a younger son, is chosen as the next leader because he defeated his brother to restore his father to the throne.

Ether 7:9 Wherefore, he came to the hill Ephraim, and he did molten out of the hill, and made swords out of steel for those whom he had drawn away with him; and after he had armed them with swords he returned to the city Nehor, and gave battle unto his brother Corihor, by which means he obtained the kingdom and restored it unto his father Kib.
Ether 7:10 And now because of the thing which Shule had done, his father bestowed upon him the kingdom; therefore he began to reign in the stead of his father.

It is understandable that Kib would reward his son, Shule, with the kingdom, since he was the one to get it back. There are situations like this. But there are other cases such as in Ether 10:4 and maybe

Ether 11:4 that seem to have no reason for the king to bestow the kingdom on a younger son.

So why would the kingdom go to a son other than the eldest? There are a few possible reasons. The first reason might be that the oldest sons may not want to rule, as previously mentioned. Nibley [5] said a king had to know that he could easily become a target for assassination or capture. We see in the crowning of the very first king, Orihah, that he was the only son of both Jared and Jared's brother that would accept the responsibility of being the king. Of course, at the time, the people were righteous, so assassination or captivity was unlikely. But even when the people are righteous, it could be an enormous responsibility being the one to judge the people.

There are other possible reasons. In researching this, the succession of kings in China gave some interesting insight, as defined in Wikipedia. [33]. Most of the Asian leaders had multiple wives. We can assume this is the case with Jaredite kings because of the number of children they often had. In the Orient, the wives usually had a status level. One wife would be the favored wife, and the next king would be chosen from her sons, usually her oldest, even if there were sons from other wives who were older. In some kingdoms, that was always done, but in others, the king would choose among his sons, and until he decided, it was not known who the king would be. We saw a similar succession in Saudi Arabia recently, with a different prince being chosen to rule than was expected.

So, the practice of having the kingdom go to a son that was not the oldest was not unusual.

8
Why is the Brother of Jared not Named?
✦

Why is the brother of Jared not named? This is one of the most often-asked questions I receive, and there could be a few different answers. But before I talk about this, it should be mentioned that we do know the name of the brother of Jared as Mahonri Moriancumer, given through modern revelation to the prophet Joseph Smith.

"While residing in Kirtland Elder Reynolds Cahoon had a son born to him. One day when President Joseph Smith was passing his door, he called the Prophet in and asked him to bless and name the baby. Joseph did so and gave the boy the name of Mahonri Moriancumer. When he had finished the blessing he laid the child on the bed, and turning to Elder Cahoon he said, the name I have given your son is the name of the brother of Jared; the Lord has just shown [or revealed] it to me. Elder William F. Cahoon, who was standing near heard the Prophet make this statement to his father; and this was the first time the name of the brother of Jared was known in the Church in this dispensation." [34].

In Ether 2:13, the place where they dwelt while building the boats that were "tight like unto a dish" was called Moriancumer. The question remains as to why his name was never recorded in the book of Ether. Before I get into that, I thought it might be good to look at the relationship between Jared and his brother.

It appears that Jared is the oldest, while his younger brother is the prophet. I suggest this because Jared is the one who makes the decisions for the people and asks his brother to go to the Lord on behalf of the people. Having an older brother leader and a younger brother prophet is not without precedence.

In Egypt, Aaron was one of the leaders of the people, and Moses was his younger brother and the prophet. Moses, of course, was called of God and became more of the leader under God's direction than Aaron. In the Book of Mormon, this type of relationship appears to hold with Ammon and Aaron, the sons of Mosiah. A person might include Laman and Nephi in this list. In modern times, Hyrum was the patriarch, and Joseph was the prophet. Again, an older brother leader and a younger brother prophet.

As I talked in class about this possible relationship, a student of mine, who had gone on a mission to Taiwan, said that there is no word in the Chinese language for brother, only older brother and younger brother. He said the Book of Mormon translation into Chinese uses older brother for the brother of Jared.

As I considered this and how it doesn't match how I feel their relationship plays out, I had a chance to meet with a man who translated the Book of Mormon into a language of a Polynesian island where he served his mission. He said that a person must be careful taking a translation as perfect and exact when it is changed to a language where a word or phrase does not exist.

As an example, he talked about 1 Nephi 11:8.

And it came to pass that the Spirit said unto me: Look! And I looked and beheld a tree; and it was like unto the tree which my father had seen; and the beauty thereof was far beyond, yea, exceeding of all beauty; and the whiteness thereof did exceed the whiteness of the driven snow.

He said snow did not exist on the island he was translating for, so there was no word for it. He didn't know what to do. He asked the brethren, and they said for him to do his best and choose something. He was told that if it was sufficient for what the Lord wanted to get across, even if it wasn't perfect, the Lord would be okay with it. If it wasn't, the spirit would dictate to the brethren that reviewed the translation, even if they couldn't speak the language, that the section containing it needed to be changed.

The man said he thought of the whitest thing he knew of on the island and used the phrase, "the whiteness of a **cockatoo feather**." He said that the brethren passed it.

So, it still could be that Jared was older, as it appears, and the translation using older brother for the brother of Jared was not of concern to the Lord. But we don't know for sure. If Jared was the older brother and leader of the people, there are some additional possible reasons I will get to later for the brother of Jared not being named.

There is one last comment I'd like to mention here. The brother of Jared was a prophet and must have had revelation ahead of the destruction of the tower of Babel and the dispersion that it would happen. The scriptures indicate that Jared asked his brother to ask the Lord that their languages not be confounded, not that they would be unconfounded. So the confounding of languages had not yet happened, and they had a foreknowledge that it would happen.

Ether Family Lineage

Going back to the fact that the brother of Jared is not named, Nibley [5] suggests that the reason might be because Ether (and possibly other record keepers) were descendants of Jared, so only his lineage is given. This is a plausible scenario. Daniel Ludlow [12](p310) says

"The Book of Ether is clearly a family record of Jared, not the brother of Jared—Ether the final writer, and perhaps the abridger, of the record — was a descendant of Jared and might naturally have emphasized the achievements of his direct ancestor rather than the brother of his ancestor."

One case that might suggest this is true is that the book a couple of times does not mention the name of a man who rose to fight the king, using the term "a mighty man" instead. In only one case was the genealogy of such a man given, and he was a descendant of the brother of Jared (Ether 11:17).

Because of such things, perhaps it was only Ether's line that was important to him. However, we have names of people given who were not Ether's progenitors, including all those who led the fight in the final battles.

In the Succession of Kings table in the appendix, there are others named whose genealogy is not given, so we don't know if they are descendants of Jared. At least in one case, a brother of a king is not named, and he would have had to have been a descendant of Jared because his brother was (Ether 10:14). But he is just called a brother of the king that he put in captivity.

As previously mentioned, there are other important people who are not named. More than once the record just says someone who came to rule was "a mighty man" and does not give their name. The inconsistency in the idea that the lack of a name was determined by lineage makes this hypothesis unlikely.

There are two other reasons that the brother of Jared's name might not have been mentioned that have been suggested by Daniel Ludlow [12] as well as a third listed in Wikipedia [35]:

1. The brother of Jared may have omitted his name out of modesty (John the Beloved did essentially the same thing in the Gospel of John, which he wrote).

2. Moroni may have omitted the name in his abridgment because of difficulty in translating (or "transliterating") the name into the Nephite language.

3. In the Roman practice of damnatio memoroae, names were intentionally removed from the record. Egyptian factions also wiped out names and statues of heretical rivals. In Ether 11:17–18, a direct descendant of the brother of Jared, and rival king, is also not named.

Name not Included out of Modesty

Regarding the first item, it seems unlikely that the brother of Jared omitted his name out of modesty. There are too many instances in the book of Ether where names have been omitted. Some of these are men lusting for power, and the probability such men would have their names omitted is close to zero. If the brother of Jared was the only one, it would be understandable. But since there are so many others, the omissions are common, so there needs to be a more common reason.

Translation Difficulty

The second reason, that it was difficult for Moroni to translate the name into the Nephite language, seems more plausible. This could account for the many names not given. If there were many challenges between the two languages, there could be multiple names that had to be referenced in another way.

I have heard it also suggested that it might be that the name was translated between Ether's language and Moroni's, but not from Moroni's to ours, and it is Joseph Smith who had to write it that way.

This is more doubtful since Joseph Smith reveals the brother of Jared's name. With English being a sound-based written language instead of word-character-based writing, this scenario is almost impossible to imagine.

Moroni's written language, based on Egyptian, was more symbol and meaning based and that would likely have caused more problems between the translations. For this reason, the challenge to translate from the Jaredite language to the Nephite language tends to be a reasonable idea for why his name was not given.

Name Removed from Record Intentionally

In the movie "The Ten Commandments," the Pharaoh says Moses's name is neither to be spoken nor written as punishment for Moses killing an Egyptian. This type of thing was fairly common. But this, the last of the three possibilities listed, seems as unlikely as the first one. Some of the wicked men who sought power could have had their names removed later by the record keepers. But this would make no sense for the brother of Jared. He was highly revered by his people and considered an honorable man who helped lead his people to a promised land. Why would someone want to remove reference to him from the record?

Of the four reasons given, there are only two that seem plausible. The first is that the record was a family record, and Ether kept more to family names. The second is that it was difficult for Moroni to translate them in Nephite written language. Apparently, it was difficult and time consuming to write in that language.

Ether 12:23 And I said unto him: Lord, the Gentiles will mock at these things, because of our weakness in writing; for Lord thou hast made us mighty in word by faith, but thou hast not made us mighty in writing; for thou hast made all this people that they could speak much, because of the Holy Ghost which thou hast given them;
Ether 12:24 And thou hast made us that we could write but little, because of the awkwardness of our hands. Behold, thou hast not made us mighty in writing like unto the brother of Jared, for thou madest him that the

things which he wrote were mighty even as thou art, unto the overpowering of man to read them.

Ether 12:25 Thou hast also made our words powerful and great, even that we cannot write them; wherefore, when we write we behold our weakness, and stumble because of the placing of our words; and I fear lest the Gentiles shall mock at our words.

The Nephite writing was based on that of Ancient Egypt, which was a hieroglyphic form with thousands of symbols.

Ether 9:32 And now, behold, we have written this record according to our knowledge, in the characters which are called among us the reformed Egyptian, being handed down and altered by us, according to our manner of speech.

This type of writing was very difficult. It did have some symbols that made sounds, but many, if not most, represented ideas [36].

Figure 33: Egyptian Hieroglyphics

It is quite conceivable that some Jaredite names were almost impossible to represent in the Nephite written language. This seems to be to be a very plausible reason why we do not have some names in the book of Ether.

But there are a couple of other reasons that I have come across why the brother of Jared might not have been named.

Custom of Deferring to the Leader

There is a taboo of naming issues related to the king or leader that was prevalent in Oriental cultures. Let me start the explanation of this by sharing some insights.

I am a mathematician and computer scientist by trade, teaching at Brigham Young University – Idaho. I was at a math conference and went to a discussion on the game of chess. I have often used analogies and concepts related to chess in my classes. But this discussion had a few unique things in it.

The presenter started with a legend that I had used many times. The story was that when the game of chess was invented, the king was so impressed that he offered up to half of his kingdom for the game. We'll come back to this idea later and why he would do that. But the story says that Chess Master, the game's creator, told the king no. Instead, he asked for a grain of rice for the first square, two grains of rice for the second square, four for the third, and so forth, with each square being twice the number of rice grains as the square before.

The story continues that the king felt dishonored to pay such a pittance and suggested, instead, one thousand gold coins for square one, two thousand coins for square two, three thousand gold coins for square four, and so forth, with sixty-four thousand gold coins for the last square. To the king's surprise, the Chess Master said he would prefer the grains of rice, so the king ordered it to be done, the value of each square delivered on successive days.

I imagine the servant bringing one grain of rice on day one and mumbling under his breath, "Idiot," because the Chess Master didn't take the gold coins. On day two, the servant probably came with two grains of rice and again mumbled, "Idiot." He probably did the same on day three with four grains of rice. But it didn't take too long for the rice to start to add up until there were baskets full, then wagons full, then it was impossible to deliver. So, let's look at the sum of the number of gold coins and the number of rice grains.

The sum of the number of gold coins is $1000 + 2000 + 3000 + \ldots + 64{,}000 = 2{,}080{,}000$ gold coins. The rice grains add up to $1 + 2 + 4 + \ldots + 2^{63} = 18{,}446{,}744{,}073{,}709{,}600{,}000$. This is basically 18 quintillion.

To get a feeling for how much rice this is, the seminar presenter asked for some volunteers. He had them fill a thimble with grains of rice. Once it was determined how many grains of rice would fit in a thimble, we used math to calculate how many thimbles were in a cup, then using normal measures, we could do cups to quarts, quarts to gallons, gallons to bushels, and so forth. We then calculated how many bushels of rice 18 quintillion would be.

Finally, the presenter showed the estimated amount of rice since history had been recorded up until then (the mid-1990s). It was still far less than 18 quintillion. He then mathematically projected rice production forward into the future. He estimated that if we added up all the rice the earth has produced through history and will produce into the future, the amount of rice needed to pay the Chess Master would not be have been grown on this planet until about 2055. Though that might have changed with modern fertilizers, the amount is staggering.

Needless to say, it didn't take long for the king to realize he had made a commitment he could not keep. There are historically two endings to this story. In one, the king was so impressed with the Chess Master that he made him second in command to himself over the whole kingdom. In the other ending, the king was afraid of the Chess Master because of his brilliance and had him put to death.

So, what does this story have to do with the brother of Jared not being named? Nothing, but it's a great story. No, seriously, it does have a reason. Why was the king willing to give up to half of his kingdom for the game of chess, or pay so much gold or rice? The answer is that it would help immortalize his name to succeeding generations.

For a king to be revered by the generations that followed, he needed to be known for some incredible feat. If they could not do one on their own, why not purchase one? This can be seen in many things, at least in math and other areas I have studied, where you read that the king made lots of great discoveries. The probability of this is next to zero since the king's time would have been taken up in ruling, and it would be almost impossible for someone to have expertise in the many varied fields. It is far more likely that the king paid those who did do the great things and then changed history to attach his name to them. Sometimes they might have killed the person, as in the second ending of the Chess Master, and then just taken credit for the great thing as their own.

Nonetheless, the kings often tried to have great things attributed to them, and they did not want anyone to get in the way of this glory. Understanding this, I began to come across material that indicated that kings did not want others to have names similar to theirs. This could be understandable if the king had done a lot to make his name synonymous with greatness. He wouldn't want someone else mistaken for him and receive part of the glory he had worked hard to obtain.

In ancient Chinese and Japanese cultures, certain restrictions were put into law regarding someone else having a name similar to the emperor or his progenitors [37]. This was both a cultural and a religious norm. A person with a name like the emperor, or his ancestors, was considered as lacking education and respect, bringing shame on himself and the emperor. This could include just writing the name of the emperor without permission. The result of violations of the law could be severe, including death for the person and their family members.

Sometimes when someone became the emperor, anyone with a similar name had to change their name, or at least remove or change important characters, to avoid the execution of the law. Some emperors would change their names when they became emperor to avoid the problem. It also became common practice for the parents of a child who could grow up to be an emperor to give the child a unique or uncommon name to avoid problems should the child become the emperor.

The emperor's name was also not supposed to be used commonly, much like Jehovah's name was forbidden to be said in ancient Israel. This caused a contradictory problem for the people. They weren't supposed to have a name similar to the emperor, but his name was not to be spoken, so sometimes it was nearly impossible for the people to know what the name was they were supposed to avoid. One final note here is that in studying ancient documents, a person's name could be changed because of these types of traditions; thus, what appears as two people in some writings could actually be the same person.

What does this have to do with the brother of Jared? The point of all of this is to show that there were certain social naming norms in ancient societies. It is possible that Jared, being the leader, was held in such respect that his brother was not named to avoid taking away from that honor. There could also have been some other similar social norms that are unknown to us which were part of their society and were the reason Jared's brother was not named.

Sumerian Naming Conventions

An even better look at social names and considerations with the Jaredites is to look at Sumerian naming conventions since the Sumerians, or related people, were likely the ones building the Tower of Babel or living in that vicinity. The Jaredites, therefore, would likely have come from that culture. Some of the names in the book of Ether are extremely Sumerian. For example, one of the most powerful Sumer city-states was Kish, and Kish is the name of a king in Ether 10.

17 And it came to pass that Corom did that which was good in the sight of the Lord all his days; and he begat many sons and daughters; and after he had seen many days he did pass away, even like unto the rest of the earth; and Kish reigned in his stead.
18 And it came to pass that Kish passed away also, and Lib reigned in his stead.

In fact, "Kish" is part of names throughout the book of Ether such as Akish and Riplikish. It can also be found in a Nephite name later, KishKumen. It is also a Bible name. In fact, King Saul was the son of a man named Kish (1 Samuel 9:1-2).

In Sumer, names were often given based on something about the person and could be a whole sentence. These names were often words compounded together and could be long,

Names could also be changed after birth to match something about the person. For example, a girl who was found as the only survivor after her village was washed away was given the name Tila meaning "The one who survived" [38]. Some of these names are found on clay-baked tablets from the Temple School of Nippura [39].

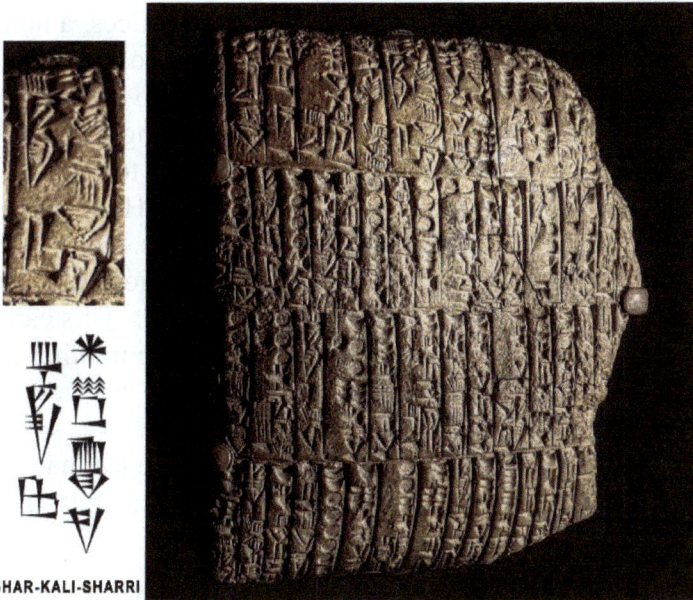

Figure 34: Example Tablet from Nippura

Examples of Sumerian names can be useful. Here a few examples of some names from Sumer that are listed along with their word meaning [40]:

Ses-tur: little brother
Ama-i-de: mother by the canal
Abzu-kurgal: the Abzu is a great mountain
Ĝakanaheti: may (this child) live for my sake
Gu-be: next to the riverbank
Guduga: one with a sweet voice
Iginisig: blue/green eyed one
Inimzida: having righteous words

The name Jared is a pre-flood name, first found in Genesis 5:18-20:

18 And Jared lived an hundred sixty and two years, and he begat Enoch:
19 And Jared lived after he begat Enoch eight hundred years, and begat sons and daughters:
20 And all the days of Jared were nine hundred sixty and two years: and he died.

The name Jared as found in the Book of Jubilees, a noncanonical book that is found in some religions and correlates to the Old Testament book of Genesis, indicates the name Jared means "to descend" (Jubilees 4:15) [41]. This could relate to the descending of angels in early Biblical writings. So, in Sumerian, we might consider Jared to mean "He Who Descends" or perhaps "He Who Came Down From Heaven." As a parent, I can imagine a mother naming her baby this.

As explained in the Sumerian name convention, a name can be given related to the person and what they have done. This can be changed later in life. This is not without precedence in other cultures. The Chinese were another culture that did this a lot. So were the Native Americans.

I had a good friend when I was young whose last name was Red Fox. Native American names and their translation can also give us some insight. A small sampling includes [42]:

For Men:
Dustu - Cherokee. Spring frog
Enapay - Sioux. Appears bravely.
Keokuk - Sauk. One who is alert and watchful.
Kosumi - Miwok. Fishes for salmon with a spear.
Seattle - Salish. Man of high status.
For Women:
Nokomis - Chippewa. Daughter of the moon.
Winona - Sioux. The first daughter.

Hawaiians still give such names today. I have some Hawaiian friends with names long enough that they are almost impossible for someone who does not speak the language to remember. I can't say the names, let alone remember them, so we use something much shorter.

In the Sumerian King List, there are names of the kings given and recorded on baked clay tablets [43]. In the list, the names are translated, along with their meaning when known. In most cases, these kings became kings after their fathers, and their name meant "Son of WhoeverTheLastKingWas." Dumu, "Son of" or "Child of" appears to be the most common part of a name. Other common things include a city, land, or something like that indicating the place the person is from. We'll see this later as we talk about Nephite names.

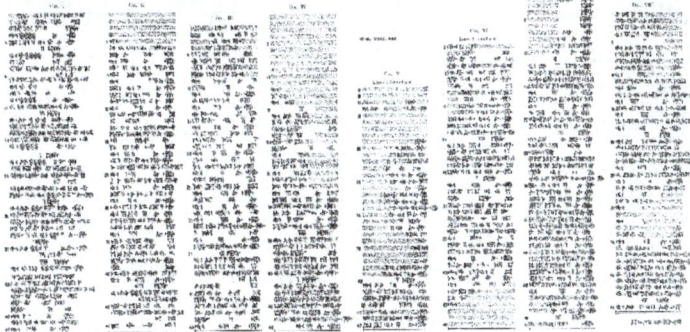

Figure 35: Sumerian List of Kings

At least in one case in the Sumerian List of Kings there is a brother reference. Manishtushu means "brother of Rimush, the son of Sargon." (His brother was the king before him.)

Where are we going with all of this? As I considered that the naming convention of the Sumerians, I realized the names of the people in the book of Ether could have a meaning. And it came to me that the name of the brother of Jared could translate just as that. His name is "The Brother of Jared" when translated. It is possible that when Moroni translated the record using the interpreters, the translation given was the name's meaning and not how the name sounds. This could mean that others whose names we considered as not given actually were given but in the form of the meaning.

For example, information on Sumer [44] talks about how a common name for a king, or at least a portion of the king's name, would be lu-gal. This means "great man" or "king" and might be the reason we find some men in the book of Ether given as "A Mighty Man" or something similar. That might be the translated name by meaning and not by sound.

Ether 11: 15 And it came to pass that there arose a rebellion among the people, because of that secret combination which was built up to get power and gain; and there arose a **mighty man** among them in iniquity, and gave battle unto Moron, in which he did overthrow the half of the kingdom; and he did maintain the half of the kingdom for many years.

Ether 11: 17 And it came to pass that there arose another **mighty man**; and he was a descendant of the brother of Jared.

In other words, maybe in wondering why the brother of Jared is not named, we (or at least I) have been looking beyond the mark. Maybe "The Brother of Jared" is his name as translated by meaning, and not phonetically, and that is the reason it is in the book of Ether that way. Perhaps Mahonri Moriancumer even translates as "The Brother of Jared" and we have therefore actually had his name in the book of Ether in its meaning form and not in its phonetic form.

I think that if Mahonri Moriancumer does translate to "The Brother of Jared" that this is only a portion of the translation. The reason for this is that in Sumerian names, Mah (as is part of the name Mahonri) has a definition of "chief, exalted, or large" [45]). It is interesting to note that the brother of Jared is called a large man. (Note the phrase "mighty man" is also here again.)

Ether 1: 34 And the brother of Jared being a **large** and **mighty man**, and a man highly favored of the Lord, Jared, his brother, said unto him: Cry unto the Lord, that he will not confound us that we may not understand our words.

It might be that the full translation by meaning of his name is something like "A Large And Might Man, The Brother of Jared" which Moroni, or the Urim and Thummim, shortened to "The Brother of Jared."

Perhaps in cases where we see the book of Ether say a "mighty man," that man's name is something like "Lu-gal" which means "A Mighty Man" or "A Great Man." That would mean we have no reason to wonder further about men referred to in this way because that is the name they are given, and the translation is in meaning form.

Even in our English language, names also have meanings, though most of us don't know them until we want to name a baby and look it up.

For example, the name "Andrew" means "power." So perhaps in our language, we could be saying, "and there arose an Andrew" instead of "and there arose a Mighty Man."

Obviously, not all names in the book of Ether are given in a form that tells us their meaning. But it seems reasonable that in the translation, some of the names could be given in their meaning form and some in phonetic form.

Conclusion - the Brother of Jared's Name

In conclusion, I feel that the Sumerian naming form and the difficulty in translating from the Jaredite writings to the Nephite written language are the two reasons that makes the most sense when answering the question as to why some people don't appear to be named in the book of Ether. It could be either of these, or a combination of both.

Personally, I feel it is the combination of these two. I feel that because the Egyptian language was so hard to write in, some Jaredite phonetic name translations might have been almost impossible for Moroni to include. It is likely that when that was the case, he instead translated it using the meaning form. It also raises a question as to whether the Urim and Thummim presented it to him that way, or if he was forced to make that decision himself. Whichever way it was, I feel some names in the book of Ether were given in phonetic form when it was feasible in the Nephite reformed Egyptian, and when it wasn't feasible, they were given in their meaning form.

9

What did Ether Mean Saying the Jaredites Were Destroyed?
Jaredites Among the Nephites

✦

This section has many things to consider, and making it cohesive and fluid, instead of choppy and segmented, is challenging. But the question that is the title of this chapter raises many more questions which can help direct the discussion here. Here are some questions:

- Geographically, where were the Jaredites in comparison to the Mulekites, Nephites, and Lamanites?
- What are the tendencies of populations during a war?
- Is there evidence of Jaredites among the Mulekites and the Nephites?

Basic Location of Jaredites, Mulekites, Nephites, and Lamanites to Each Other

There is much debate about where the Jaredites, Mulekites, Nephites, and Lamanites were located. That is not our discussion here. All we need to understand is that, basically, the Jaredites were north of the Mulekites, the Mulekites were north of the Nephites, the Nephites were north of the Lamanites, and there was lots of wilderness around all of them.

Ether 9:31 And there came forth poisonous serpents also upon the face of the land, and did poison many people. And it came to pass that their flocks began to flee before the poisonous serpents, towards the land southward, which was called by the Nephites Zarahemla.
Ether 10: 19 And it came to pass that Lib also did that which was good in the sight of the Lord. And in the days of Lib the poisonous serpents were destroyed. Wherefore they did go into the land southward, to hunt food for the people of the land, for the land was covered with animals of the forest. And Lib also himself became a great hunter.

Ether 10: 20 And they built a great city by the narrow neck of land, by the place where the sea divides the land.

Ether 10: 21 And they did preserve the land southward for a wilderness, to get game. And the whole face of the land northward was covered with inhabitants.

See also Alma 22: 27-34 for Mormon's description of how the people were located in the land and how the lands were laid out.

This is important because we need to know the general location of the different people to each other so we can understand how they could have intermingled. For example, with the Jaredites north of the Mulekites and the Nephites south of the Mulekites, it might be easier for the Jaredites to have had contact with the Mulekites than with the Nephites or Lamanites. However, there was wilderness around all of them, which could have made it possible for the groups to have contact with each other without going through a particular group.

In a simplified version of the land, over time, we had something like this:

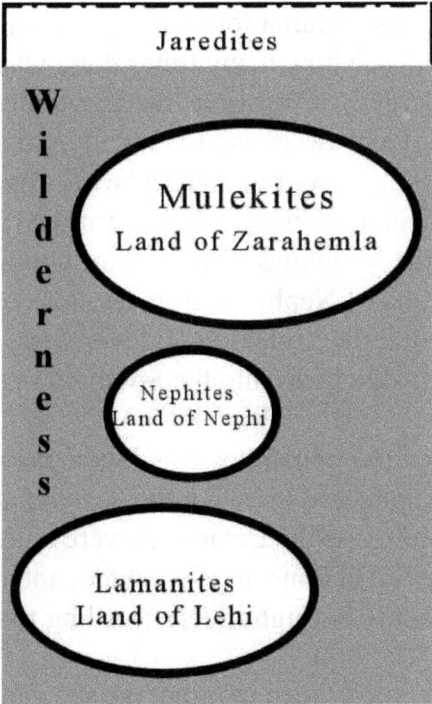

Figure 36: Jaredites, Mulekites, Nephites, and Lamanites in Relationship to Each Other

When Lehi's family first came to America, all his people were together near the place they landed, which was called "The land of Lehi." The Jaredites still existed as a nation, though they might have been moving into the final stages of their destruction.

Lehi left Jerusalem in the first year of the reign of Zedekiah (1 Nephi 1). Zedekiah reigned for about ten or eleven years, then Jerusalem was destroyed. The Mulekites came after that destruction. With the eight years Lehi's family spent in the wilderness, and the time in Bountiful building the ship to come to the promised land, it is possible the Mulekites could have arrived in the promised land near the time Lehi did. The Mulekites settled north of the Nephites and Lamanites.

Not too long after getting to the promised lands, Nephi's brothers were threatening to kill him. So, Nephi took those who would follow him, left the Lamanites, and formed the Nephite nation in the Land of Nephi as seen in Figure 36: Jaredites, Mulekites, Nephites, and Lamanites in Relationship to Each Other.

The gray in Figure 36 represents the fact that a lot of wilderness surrounded these groups. Search for "wilderness" in the Book of Mormon, and you will see how many times there was travel or related instances in the wilderness. Over time, the Lamanites gradually took over most of these wilderness areas. Eventually, after a couple of hundred years, the Nephites were about to be destroyed by the Lamanites. King Mosiah the first took the Nephites who would follow him and left to join the Mulekites (The book of Omni). The Lamanites then took over the Land of Nephi. The Lamanites continued moving into wilderness areas around the Mulekites and Nephites, and eventually, the Nephites worked to hedge them into the Southern regions only.

Alma 22:34 Therefore the Lamanites could have no more possessions only in the land of Nephi, and the wilderness round about. Now this was wisdom in the Nephites—as the Lamanites were an enemy to them, they would not suffer their afflictions on every hand, and also that they might have a country whither they might flee, according to their desires.

With this geography idea in mind, we see that the Jaredites were closest to the Mulekites and likely had the most interaction with them. However, with large swaths of wilderness, it would be possible for them to interact with all three groups, Nephites, Mulekites, and Lamanites, without crossing through the others. With these geographical ideas laid out, we can move to other related concepts.

Tendencies of Populations During War

Nibley [5] says that when war breaks out, people flee the areas where the fighting occurs. We see this in the wars that are going on in the world today. Wherever there is war, there is a massive migration of war refugees. Nibley says it is unreasonable to think that the same thing would not have happened with the Jaredites as long as they had somewhere to go.

There was mainly only one way they could flee, and that was south. There is a possibility that there was wilderness or some other protection to the north, maybe even the Arctic regions. But if they fled north, they could only go so far before the land would become so inhospitable that it could not support them in large numbers. So, it would be far more likely that they would head south if they were to flee the fighting.

As the war continued, the leaders of the two sides went about gathering all they could who would join them. Those who would not join them were being slain.

Ether 15:14 Wherefore, they were for the space of four years gathering together the people, that they might get all who were upon the face of the land, and that they might receive all the strength which it was possible that they could receive.

It is impossible to believe that there weren't people fleeing from the war zone during this gathering for battle. Surely some who didn't want to fight would have done what they had to do to save their families and themselves from this indescribable horror. It says they gathered all that were in the land, but many might have left. That brings us to the next question. Is there evidence of Jaredites among the Mulekites, Nephites, and even among the Lamanites?

Evidence of Jaredites among the Mulekites and the Nephites

There are a few things we can look at to answer this question.

1. Are there Jaredite names among the Mulekites and Nephites?
2. Nephite weights and measures
3. Are there places among the Mulekites that seem to relate to the Jaredites?
4. Other scriptural clues

Jaredites Among the Nephites

One of the best ways to determine if there were people fleeing from the Jaredite destruction to live among the Mulekites and Nephites is to see if any of the names appear in both places. This would include names taken from the same root with various changes.

Not including Bible names in Isaiah references, there are around 375 names in the Book of Mormon. This includes land and city names. I started to put them in myself, an arduous task, but then considered that someone had probably done that. I found a list of place names at https://en.wikipedia.org/wiki/List_of_Book_of_Mormon_places. A list of names of the people can be found at https://en.wikipedia.org/wiki/List_of_Book_of_Mormon_people.

I put these lists into a spreadsheet and did a fair amount of editing. Nibley [5] says that no Jaredite names appeared in the Nephite record until Mosiah the first took his people to join the Mulekites. I'm afraid I have to disagree with that, and I will explain why later. But using that assumption, I marked whether the names that came from the Jaredite record (Ether), were before Mosiah in the Nephite record, or were after Mosiah in the Nephite record. I then sorted them by name. I used that to find similarities.

I created a table showing Jaredite names and Nephite/Mulekite names that could be derived from them. Considerations from the derivations will be discussed later. For both Nephite/Mulekite names, I will use "Nephites" in the explanations. In the middle column, I used B = Before Mosiah and A = After Mosiah to help determine the break between where the two groups joined. Both letters were used if the name(s) appeared in both parts of the record.

Table 2 - Names Common or Similar Between Jaredites and Nephites

Jaredite Name		Nephite/Mulekite Name
Aaron	A	**Aaron** – found both among the Nephites and the Lamanites.
Ahah	A	**Aha** – Nephite military officer **Ammah** – Nephite missionary
Kish, Akish, Riplakish	A	**KishKumen** – Gadianton robber leader
Moron	B A	**Moroni** – 2 great Nephite leaders, **Moronihah** – Nephite leader **Amaron** – son of Omni **Amoron** – commander in Mormon's army **Ammoron** – Nephite dissenter
Cohor, Corihor	A	**Korihor** – antichrist
Coriantor	A	**Corianton** – Alma's son
Coriantum, Coriantumr	A	**Coriantumr** – Descendent of Zarahemla and commander of Lamanite forces
Emer	A	**Emron** – Nephite soldier
Gilgah, Gilgal	A	**Gilgal** – Nephite commander, land and city name
Jared	B	**Jarom** – Great grandson of Lehi
Morianton	A	**Morianton** – founder of city and city name **Gadianton** - wicked robber chief (one I missed and added after reading Nibley work pointing to the similarities)
Nehor	A	**Nehor** – Nephite apostate Also, a land in both cases, and perhaps the same land.
Noah	A	**Noah** – wicked Nephite king
Omer	A B	**Omner** – Mosiah's son, also city by the sea **Omni** – record keeper son of Jarom
Orihah	A	**Onidah** – land
Riplakish	A	**Riplah** – name of hill
Ramah	A	**Cumorah** – similarities of names between where the final battles of the Nephites and Jaredites occurred. Moroni's words seem to indicate they are the same place. (There is more discussion on this below.)
Shared	B	**Sherem** – antichrist
Shez	A	**Shem** – Nephite commander and city name

Shiblom, Shiblon	A	**Shiblom** – Nephite commander, **Shiblon** – son of Alma, **Shilom** – city name
Shim	A	**Shim** – hill name for both Jaredite and Nephite people. Likely the same hill, the name being carried from Jaredites to Nephites. **Shilom, Shemlon, Shimnilon** – all cities of Lamanites named together as cities of Mosiah's sons' missionary conversions.
Mahonri Moriancumer*	A	**Moroni** – Nephite general, also great prophet **Coriantumr** – Commander of Lamanite forces
Other Possible Nephite Names that Could Have Jaredite Basis		
Chemish – grandson of Omni. **Helam, Helaman,** maybe **Helorum** – leaders of the Nephites.		
*Mahonri Moriancumer – the name that Joseph Smith revealed as the name of the Brother of Jared, is not found in the book of Ether, except part of it in the name of a land. But the name has similarities to other names that are in The Book of Mormon.		

After I finished my analysis, I compared it to Nibley's [5]. How I did my analysis was different from his. I looked at commonality of names, including roots with changes or additions, as described in the section on Sumerian names in the part of the book about why the brother of Jared might not be named. (If you haven't read it, go back and read it. It adds a lot here.)

Nibley [5] did his analysis based on the names themselves. He says that the Jaredite names often had an -m ending (mimation) and later an -n ending (nunation), like in curelom and cumom for animals and shalom for a mountain. He also points out that many of those with what appears to be Jaredite-affiliated names were rebellious against the Nephites. Though he does not directly list all the name similarities, he specifically defines a few. Because of the tendency to be against the Nephites, he adds the name Gadianton to the list since it is close to Morianton.

It might be worth analyzing the different names more. One name, Aaron, is common enough that a person might expect to find it in both records, even if there was no crossover. There is also one that is common to both, which might still be because of Jaredite influence. Nibley [5] points out that Noah is found in the Jaredite record and only once in the Nephite record. Noah was the son of Zeniff, a person who was likely of Mulekite descent. Nibley also points out that Noah is not even a common name in the Bible, only used for the man who built the ark.

Most of the other names are not common. Nibley [5] says there is a Mulekite connection to all the names that are a crossover between the Nephite and Jaredite records. However, from the common names table above, you may note that I show a couple of exceptions. Those are the ones marked with the B in column two. But before I talk about why, let's consider other things about the names themselves.

Nibley [5] mentions that the first place the Jaredites settled in this land was Moron. Adding the -i ending to a word in ancient Egyptian gave it ownership, so Moroni would mean belonging to Moron. There are other names in the table that have a connection to Moron using other word additions. An example is Ammoron. The -am- prefix comes from Latin and has two meanings, love or to come out of [46]. It is possible that when Ammoron was born, his mother named him that to show her love. But it is more likely that it shows he is from Moron.

This raises an interesting question. As you may recall, Ammoron succeeded Amalickiah as king after Amalickiah was killed leading the Lamanite armies against the Nephites. Ammoron then continued the war. Captain Moroni commanded the armies for the Nephites, and there seemed to be some underlying animosity between them, as seen in their letter exchange (Alma 54). The -i ending on Moroni's name shows he was probably from the land of Moron. The -am- prefix on Ammoron's name likely means the same thing for Ammoron. This makes a person wonder if they might have known one other before they ever came to war against each other.

One other note here about the -am- prefix. The Sumerian word combinations mentioned in the section about why the brother of Jared is not named would have been known and would have been part of the culture of the Jaredites. However, Latin additions and endings would not have been known to the Jaredites, but likely would have been known to the Nephites.

Latin influence began around the Mediterranean Sea about a hundred years before Lehi. Latin influence in Judah could easily have been spread there by the Phoenicians, who were the sea merchants of the Mediterranean and beyond. By the time Lehi and the Mulekites came to the Americas, the Latin influence could have been part of their culture, even though it would have come along far after the Jaredites. So, a name like Ammoron, though a possible derivative of the Jaredite name, Moron, would not have been familiar to the Jaredites, but reasonably could have been to the Nephites. This could be true of other names among the Nephites that have Jaredite roots.

Another thing to consider is how names change from generation to generation. An example of this is some women I know. The first lady in the line was named Eliza. She named her daughter Lillian, which she considered to be a more modern form for Eliza at the time. Then, years later, when Lillian had a daughter, she named her Lilly, which was the contemporary form of Lillian then. This shows that names can process through a transition and still be from the same base.

As I look at the name table, one example I feel could be this way is Omer, to Omner, to Omni. Of course, there might not be any connection at all, but there could also be even more connections than we may notice.

Another point of interest might be in the Jaredite land of Ramah, which Moroni equates to the land of Cumorah. Moroni indicates they are the same hill.

Mormon 6:2 And I, Mormon, wrote an epistle unto the king of the Lamanites, and desired of him that he would grant unto us that we might gather together our people unto the land of Cumorah, by a hill which was called Cumorah, and there we could give them battle.

Mormon 6:6 And it came to pass that when we had gathered in all our people in one to the land of Cumorah, behold I, Mormon, began to be old; and knowing it to be the last struggle of my people, and having been commanded of the Lord that I should not suffer the records which had been handed down by our fathers, which were sacred, to fall into the hands of the Lamanites, (for the Lamanites would destroy them) therefore I made this record out of the plates of Nephi, and hid up in the hill Cumorah all the records which had been entrusted to me by the hand of the Lord, save it were these few plates which I gave unto my son Moroni.

Now the connection as given by Moroni.

Ether 15:11 And it came to pass that the army of Coriantumr did pitch their tents by the hill Ramah; and it was that same hill where my father Mormon did hide up the records unto the Lord, which were sacred.

The names Ramah and Cumorah have many similarities. The -mah- part is a term in Sumerian, meaning great, large, or magnificent. This could mean that the place the Jaredites gathered for their final battle was a magnificent place. The -co-, -com-, -cum- of the Nephite name are Latin prefixes meaning together. To the Nephites, the combined word probably meant the large or magnificent place of gathering, or coming together.

These Latin prefixes could have been added to the Jaredite word because the Nephites knew it was the final place the Jaredites gathered to battle. Since the name is attached to the surrounding lands as well as the hill, it likely had that name long before the final Nephite-Lamanite battle there. It seems ironic that the hill, with that name, would also become the gathering place for the final Nephite-Lamanite battle.

There are other names I have added to the list that likely have a Jaredite basis, though neither they, nor a related form, are found in the book of Ether. This is because they are very close to city or land names connected to the Jaredites. The relationship of the root Kish to the city of Kish where the Jaredites likely came from has already been discussed. But another city that the Jaredites probably knew about was the city of Carchemish.

Carchemish was a Bible land on the west bank of the Euphrates River [47]. At the time of the Jaredites, it would have been an important city. It was also near the city of Kish, and Sumerian articles have been found in archaeological digs at Carchemish.

Carchemish could also have been known to Lehi's people and the Mulekites, but the Jaredites would be more likely to have used names similar to it since the building of the Tower of Babel was closer. And the Sumerians, of which the Jaredites were likely part, probably had direct correspondence with Carchemish.

Chemish was a grandson of Omni, whose name also appears in the table of possible Jaredite-related names. Chemish's name is so close to the city's name that it likely has reference to it.

Another set of possibly related Jaredite names related to land areas known by the Jaredites are Helam (Alma 1 convert) and Helaman (and maybe Helorum), sons of King Benjamin. Helaman also appears many times later in the Nephite record. As previously mentioned, the city-state where the Jaredites were likely from was Kish. Kish was close to the land of Elam. At least the name Helam probably has some connection to the name Elam and shows a Jaredite influence.

Jaredite Influence Among Nephites - Before Nephites Joined Mulekites

As I have mentioned, Nibley felt there was no Jaredite influence shown in the record of the Nephites until they joined with the Mulekites.

"There is nothing in the Book of Mormon that shows *direct* contact between the Nephites and the Jaredites. There is always a go-between—the Mulekites, who, as the story of the elder Coriantumr shows, were the nearest neighbors to the Jaredites and separated, as we learn from Mosiah's accunt[sic], by a considerable distance from the Nephites. Everything points to the absorption of a good deal of Jaredite culture by the people of Zarahemla shortly after their arrival: The tradition of a very Jaredite pattern of behavior and dissent against Nephite rule by men of Mulekite background bearing Jaredite names makes the case pretty clear." [5](p246)

I usually agree with Nibley on things, but I'm afraid I have to disagree that there is nothing to show a direct connection between the Jaredites and the Nephites. I found five names in the Nephite record before they join with the Mulekites that I feel have Jaredite influence. There are more name connections after the Nephites joined with the Mulekites, but considering the record before the nations joined is shorter than the record after, it is not surprising. In comparison, there are twenty-two names total (not counting Bible names) before the two nations joined. In contrast, there are two-hundred twenty-one names after the nations combined. That is just over ten times the number of names, so more similarities would be expected after the nations joined simply due to the higher name volume.

From the name table,

Table 2, I have listed Amaron, Jarem, Omni, Chemish, and Sherem as names mentioned with a possible Jaredite connection before the Nephites joined with the Mulekites. The story of Sherem especially has some insightful pieces to it. But some of the pieces raise more questions than they answer.

Sometimes I feel we get so caught up in Sherem's apostate preaching that we miss some interesting details. Sherem has a very Jaredite-type name, both from Nibley's analysis of the -m ending and my analysis of name similarity to Shared. So, let's delve a little deeper into the story and try to pick up details, starting right at the beginning.

First, remember that Sherem meets with Jacob, the brother of Nephi, so it isn't long after Lehi's people came to the promised land. This is somewhere between 524 and 421 B.C. Lehi's party came to the promised land around 591-589 B.C. This means the Nephites would have been in the promised land between 65 and 170 years. That is a wide range, but at least we know they had been there for some time.

Jacob 7:1 And now it came to pass after some years had passed away, there came a man among the people of Nephi, whose name was Sherem.

This shows that Sherem was not part of the people of Nephi because he came among them. One idea to consider here is that the Nephite nation was likely still relatively small. It is possible Jacob, especially being a leader, knew everyone. He would be acutely aware if a stranger came among them. It can be assumed that Sherem was not a Lamanite, or he would likely not have received such a warm reception from the Nephites but would have been viewed with suspicion. So where was he from? Let's continue.

Jacob 7:4 And he was learned, that he had a perfect knowledge of the language of the people; wherefore, he could use much flattery, and much power of speech, according to the power of the devil.

This, again, is an indication that he was not a Nephite. A person would naturally assume a Nephite had a good command of the Nephite language. It might be assumed that Jacob was saying Sherem had an extensive vocabulary, but I doubt it. It seems that Jacob is saying this stranger knew the language, customs, and social norms well enough that

113

he understood what would get the people to listen to his message. But if he was of Jaredite heritage, where would he have learned so much about the Nephites? Sharem shows some of this knowledge as he speaks to Jacob.

Jacob 7:7 And ye have led away much of this people that they pervert the right way of God, and keep not the law of Moses which is the right way; and convert the law of Moses into the worship of a being which ye say shall come many hundred years hence. And now behold, I, Sherem, declare unto you that this is blasphemy; for no man knoweth of such things; for he cannot tell of things to come. And after this manner did Sherem contend against me.

This raises another question. If Sherem's name had Jaredite heritage, and he believed in Moses, where did he gain his understanding of Moses? The Jaredites left for the promised land long before the Israelites even existed, let alone Moses. There is probably only one logical conclusion: he had been among the Mulekites. He may even have been born among the Mulekites but had Jaredite heritage.

At this point, since the Mulekites and the Nephites came out from Jerusalem at nearly the same time, they probably still shared the same language, customs, and social norms. Contrast that to when Mosiah takes the last of the Nephites from the land of Nephi to join the Mulekites at probably around 150 B.C. By then, the Nephites and Mulekites had been in the promised land for around 440 years. (Note that the short little book of Omni covers 193 years, almost as much as all the rest of the Nephite record to that point.) When Mosiah's people joined those in Zarahemla, the language of the Mulekites was so different from the Nephites that they couldn't even understand each other. They also had lost all belief in God because they had brought no records (scriptures) with them.

Continuing with Jacob and Sherem, we get a scripture that raises one of the greatest questions of all.

Jacob: 7:10 And I said unto him: Believest thou the scriptures? And he said, Yea.

If Sherem was a foreigner to the Nephites yet understood the scriptures, where did he get that understanding if no one else had scriptures? I feel he learned from the Mulekites, who, at that time, still had oral traditions and teachings that were gradually lost over the years.

So, where does all of this take us? Sherem appears to be a man with Jaredite influence in his name, yet he could speak the Nephite language and knew of Moses. He almost surely had to have been among the Mulekites. This indicates that there was much more interconnect between the Jaredites, Mulekites, and Nephites than we might believe.

When looking at the other names from the table of names related to the Jaredites, if the table has any accuracy at all, there is another important consideration. If there were 5 out of 22 names that show Jaredite influence before the Nephites joined the Mulekites, that is 23%. 32 out of 221 names show Jaredite influence after the Nephites joined with the Mulekites, which is 14%. This is not really a fair comparison, however, since, in the longer record, the same names appear many times, and I only counted them each once. But it indicates that the likelihood of Jaredite influence might have been more than expected among the Nephites before they joined the Mulekites.

So how could this be that Jaredites were reaching the Nephites if the Mulekites were between them? As previously mentioned and shown in Figure 36: Jaredites, Mulekites, Nephites, and Lamanites in Relationship to Each Other, Figure 36, there was lots of wilderness areas all around the Nephites, Lamanites, and Mulekites. If the Jaredites were fleeing war, they probably scattered in many directions.

One other note here. The people of Limhi, in searching for Zarahemla, got lost in the wilderness and made it all the way to where the Jaredites had been destroyed without running into the Mulekites. Let's start with Ammon, leading the group trying to find those who went up to the land of Nephi from Zarahemla.

Mosiah 7:4 And now, they knew not the course they should travel in the wilderness to go up to the land of Lehi-Nephi; therefore they wandered many days in the wilderness, even forty days did they wander.

Now consider the people enslaved by the Lamanites in the land of Nephi sending a search group to try to find Zarahemla.

Mosiah 8:7 And the king said unto him: Being grieved for the afflictions of my people, I caused that forty and three of my people should take a journey into the wilderness, that thereby they might find the land of Zarahemla, that we might appeal unto our brethren to deliver us out of bondage.

Mosiah 8:8 And they were lost in the wilderness for the space of many days, yet they were diligent, and found not the land of Zarahemla but returned to this land, having traveled in a land among many waters, having discovered a land which was covered with bones of men, and of beasts, and was also covered with ruins of buildings of every kind, having discovered a land which had been peopled with a people who were as numerous as the hosts of Israel.

Mosiah 8:9 And for a testimony that the things that they had said are true they have brought twenty-four plates which are filled with engravings, and they are of pure gold.

Mosiah 8:10 And behold, also, they have brought breastplates, which are large, and they are of brass and of copper, and are perfectly sound.

Mosiah 8:11 And again, they have brought swords, the hilts thereof have perished, and the blades thereof were cankered with rust; and there is no one in the land that is able to interpret the language or the engravings that are on the plates. Therefore I said unto thee: Canst thou translate?

Mosiah 8:12 And I say unto thee again: Knowest thou of any one that can translate? For I am desirous that these records should be translated into our language; for, perhaps, they will give us a knowledge of a remnant of the people who have been destroyed, from whence these records came; or, perhaps, they will give us a knowledge of this very people who have been destroyed; and I am desirous to know the cause of their destruction.

Ammon says he cannot translate, but King Mosiah can. Later, in chapter 21, after the story is told of the people who had gone to the land of Nephi, it is reiterated that those Limhi sent to search for the land of Zarahemla had gotten lost and instead found the land with the destruction of the Jaredites. Starting out with Limhi's joy of finding Ammon and his brethren were from Zarahemla we read:

Mosiah 21: 24 But when he found that they were not, but that they were his brethren, and had come from the land of Zarahemla, he was filled with exceedingly great joy.

25 Now king Limhi had sent, previous to the coming of Ammon, a small number of men to search for the land of Zarahemla; but they could not find it, and they were lost in the wilderness.

26 Nevertheless, they did find a land which had been peopled; yea, a land which was covered with dry bones; yea, a land which had been peopled and which had been destroyed; and they, having supposed it to be the land of Zarahemla, returned to the land of Nephi, having arrived in the borders of the land not many days before the coming of Ammon.

27 And they brought a record with them, even a record of the people whose bones they had found; and it was engraven on plates of ore.

The key point of all of this is that if Limhi's people could get lost in the wilderness and make it all the way to the land of the Jaredites, than it is reasonable to believe Jaredites fleeing the war and destruction could flee into the wilderness and make it all the way to the Nephites and even beyond.

Notice that there appears to be Jaredite influence even within the Lamanites. The Lamanite cities of Shilom, Shemlon, and Shimnilon are examples. Of course, these might have been in the land of Nephi and might have originally received those names from the Nephites while they still occupied that land.

However, we have almost no sample of names to draw from with respect to the Lamanites. So, there could have been a huge influence of the Jaredites among them, and we just do not have any information based on names to show it. Since the Lamanites settled into much of the wilderness areas, it is likely there were a fair number of Jaredites fleeing the war of their people that ended up among them.

The main conclusion we can draw from all of this is that the similarities of the Jaredite names to many Nephite, Mulekite, and even Lamanite names indicate that the Jaredites were intermingling with these other nations.

Nephite Weights and Measures

Nibley [5] points out that the weights and measures of the Nephites have Jaredite names.

Alma 11:4 Now these are the names of the different pieces of their gold, and of their silver, according to their value. And the names are given by the Nephites, for they did not reckon after the manner of the Jews who were at Jerusalem; neither did they measure after the manner of the Jews; but they altered their reckoning and their measure, according to the minds and the circumstances of the people, in every generation, until the reign of the judges, they having been established by king Mosiah.

This scripture has two essential parts. The reckoning was not according to the Jews, and it was established by Mosiah, a king over the Mulekite and Nephite combined nation. The terms used for weight and measure are as follows (Alma 11:5-19):

- senine
- seon
- shum
- limnah
- senum
- amnor
- ezrom
- onti
- shiblon
- shiblum
- leah
- antion

These words, or at least most of them, do seem to have a Jaredite feel when compared to the names in the name table,
Table 2. This implies something interesting. The Jaredite influence among the Nephites and/or Mulekites was likely right from the beginning when the Nephites and Mulekites came to the promised land. If a nation was well established with its own weight and measure reckoning, it would be unlikely that the people would want to change, though it could happen.

An example of this was when the United States and Canada decided to go to the metric system over a three-year period in the 1970s. I can remember, as a teenager, how all the highway signs had both miles per hour and kilometers per hour. The car speedometers from that period had both as well. When the three years were up, Canada pushed on to make the change. However, in the United States, there was such a fuss put up by the older generations that were used to the English-based system that congress backed off, and the United States never made the change. For the change to ever happen, we would probably need forty years in the wilderness for the older generation to pass away and for the new generation to get used to the change.

Anyway, the weights and measures having a Jaredite flair to them likely indicate that there was Jaredite influence right from the beginning when the Nephites and Mulekites came. Of course, it could be like with Canada, and the people just made the changeover.

Interconnect of Place Names Between Nephites and Jaredites

There is a strong suggestion in the Book of Mormon that names given to places in the Jaredite record were known to the Nephites. Sometimes these places have the same name among the Nephites as they did among the Jaredites. That would be a strong sign that Jaredites mingled among the Nephites because it could not have happened otherwise.

An example of the name being the same is with the hill Shim. When Mormon is ten years old, Ammaron, an old Nephite record keeper, tells him about the hill Shim.

Mormon 1:3 Therefore, when ye are about twenty and four years old I would that ye should remember the things that ye have observed concerning this people; and when ye are of that age go to the land Antum, unto a hill which shall be called Shim; and there have I deposited unto the Lord all the sacred engravings concerning this people.

Now compare that to the record of the same hill as written by Moroni from the book of Ether.

Ether 9:3 And the Lord warned Omer in a dream that he should depart out of the land; wherefore Omer departed out of the land with his family, and traveled many days, and came over and passed by the hill of Shim, and came over by the place where the Nephites were destroyed, and from thence eastward, and came to a place which was called Ablom, by the seashore, and there he pitched his tent, and also his sons and his daughters, and all his household, save it were Jared and his family.

119

Not only does Moroni call the hill Shim, but he seems to know exactly where it is compared to Nephite landmarks. It is doubtful that if he was strictly going by Ether's descriptions in the record that he could place them in correlation to the landmarks known to the Nephites. This is not the only time in the translation of the book of Ether that Moroni shares an event in the book of Ether and gives a corresponding Nephite geography and event. He does it a couple of other times as well.

Ether 7:6 Now the land of Moron, where the king dwelt, was near the land which is called Desolation by the Nephites.

Ether 15:11 And it came to pass that the army of Coriantumr did pitch their tents by the hill Ramah; and it was that same hill where my father Mormon did hide up the records unto the Lord, which were sacred.

My feeling is that these names and places, named in the Jaredite record, had to be known among the Nephites. It is hard to imagine that Ether could put enough description in his record for Moroni to have known these places just by their description. Mormon has given an abundance of geographical description in The Book of Mormon, and still, people debate where those places are. It seems plausible that the only explanation for Moroni to know them is that there were Jaredites living among the Nephites who brought the information with them and shared it with the Nephites.

Other Scriptural Clues of Jaredites Among the Nephites

There is another scriptural clue that there were Jaredites that had intermingled with the Nephites. It appears at the point that Limhi's people have found the twenty-four gold plates of Ether and are eager to know what they say. Ammon was the leader of the group sent from Zarahemla to check on this people. Limhi inquired of Ammon if he knew anyone who could do the translation of the twenty-four gold plates.

Mosiah 8:12 And I say unto thee again: Knowest thou of any one that can translate? For I am desirous that these records should be translated into our language; for, perhaps, they will give us a knowledge of a **remnant** of the people who have been destroyed, from whence these records came; or, perhaps, they will give us a knowledge of this very people who have been destroyed; and I am desirous to know the cause of their destruction.

It is interesting to note the word "remnant" used here by Limhi. A remnant is not something that is destroyed, but a part which is left after something is destroyed. Limhi appears to be aware of those who are a remnant of the people who were destroyed. So why are the people so eager to know the story that is on the plates if they have people among them who were descendants of those destroyed? Speaking of Mosiah doing the translation of the twenty-four gold plates, The Book of Mormon says:

Mosiah 28:12 And this he did because of the great anxiety of his people; for they were desirous beyond measure to know concerning those people who had been destroyed.

Mosiah 28:17 Now after Mosiah had finished translating these records, behold, it gave an account of the people who were destroyed, from the time that they were destroyed back to the building of the great tower, at the time the Lord confounded the language of the people and they were scattered abroad upon the face of all the earth, yea, and even from that time back until the creation of Adam.
Mosiah 28:18 Now this account did cause the people of Mosiah to mourn exceedingly, yea, they were filled with sorrow; nevertheless it gave them much knowledge, in the which they did rejoice.

A person might ask, how could the people not know the story of the Jaredites if they had those living among them who were their descendants? It would be due to the passage of time. This translation was done around 92 B.C. The Nephites came around 600 B.C. The Nephites and Mulekites probably overlapped the Jaredites by 50 to 250 years. So, at a minimum, it has been around 250 years since the Jaredite's destruction, and it was likely much longer. And as noted about the Mulekites, they apparently didn't keep records because their language changed to the point that Mosiah and his people couldn't even understand them.

The people probably knew of the destruction of the nation to the north. But it is likely that the stories of where they came from and details about their lives were lost to time. This record would have been especially of interest to those who were descendants of the Jaredites.

Conclusion of Jaredites Among the Nephites

There is a lot of evidence that Jaredites existed among the Nephites, probably fleeing the destruction of their own nation. There are name similarities, and the weights and measures have a Jaredite feel to them. There is also an indication that Moroni knows geographical places named in the book of Ether. But didn't Ether say they were destroyed?

Nibley [5] points out that they were destroyed as a nation, not necessarily annihilated as a people. He reminds us that Ether was still alive after writing the record. The Lord has said many times that he would destroy a people, including Israel and Judah, but descendants of these people still exist. We also know that there were Nephite dissenters among the surviving Lamanites.

Mormon 6:15 And it came to pass that there were ten more who did fall by the sword, with their ten thousand each; yea, even all my people, save it were those twenty and four who were with me, and also a few who had escaped into the south countries, and a few who had deserted over unto the Lamanites, had fallen; and their flesh, and bones, and blood lay upon the face of the earth, being left by the hands of those who slew them to molder upon the land, and to crumble and to return to their mother earth.

In fact, at the time of Christ, the Nephites and Lamanites combined.

4 Nephi 1:17 There were no robbers, nor murderers, neither were there Lamanites, nor any manner of -ites; but they were in one, the children of Christ, and heirs to the kingdom of God.

But later, they split according to belief, not linage (4 Nephi 1:36-37).

4 Nephi 1:36 And it came to pass that in this year there arose a people who were called the Nephites, and they were true believers in Christ; and among them there were those who were called by the Lamanites—Jacobites, and Josephites, and Zoramites;

4 Nephi 1:37 Therefore the true believers in Christ, and the true worshipers of Christ, (among whom were the three disciples of Jesus who should tarry) were called Nephites, and Jacobites, and Josephites, and Zoramites.

4 Nephi 1:38 And it came to pass that they who rejected the gospel were called Lamanites, and Lemuelites, and Ishmaelites; and they did not dwindle in unbelief, but they did willfully rebel against the gospel of Christ; and they did teach their children that they should not believe, even as their fathers, from the beginning, did dwindle.

4 Nephi 1:39 And it was because of the wickedness and abomination of their fathers, even as it was in the beginning. And they were taught to hate the children of God, even as the Lamanites were taught to hate the children of Nephi from the beginning.

So even the Nephites were not totally destroyed physically. What God destroyed was their national identity. Nibley shares this idea:

"The dropping of the name Jaredites by their mixed descendants has many historical parallels. Thus the Hurrians lost their name so quickly and completely when they mixed with the Hittites that until recent years it was doubted that there ever were such people; yet we now know that it was the Hurrians, ranging over the vast back country to the north, that supplied the Hittites with their ruling class and their tradition of empire. Such a role may the scattered and nomad Jaredites of the last days have played in contact with the more civilized but less aggressive people of Zarahemla, completely losing their Jaredite identity but still given away, as are the Hurrians, by the strange names of their leaders." [5](p246)

It is plausible that many people of Jaredite heritage lived among the Nephites, Mulekites, and Lamanites. So why doesn't the record say something about them among the Nephites?

The Book of Mormon only has three verses in Omni (Omni 15-17) that mention the Mulekites. The book of Omni is on the small plates of Nephi, which Mormon didn't even plan to include in the record. But the Lord inspired him to add them, knowing the 116 pages would be lost. If he hadn't, we might not have had anything about the Mulekites.

It appears that writing about other nations besides the Nephites and Lamanites was outside Mormon's purpose in compiling the record.

10
Are there ideas about what the Jaredites might look like?
✦

I had just finished an Education Week class on the Jaredites when a lady approached me.

"Brother Howard, I have a question," she said. "I loved your presentation, but it left me with a big question. Do you have any idea what the Jaredites looked like?"

"Well, we know that Moroni used the word 'fair' to describe the people," I replied.

"True," the woman said. "But if they came through the Orient, and possibly left people there as you suggested, would that mean the Jaredites were of an oriental complexion? Would those people be considered fair?"

I couldn't answer either of those questions. "Fair" might mean something different to different people. It might mean light skin and hair, or it could mean that they were pretty or handsome.

I had never considered the question of what the Jaredites might have looked like before, and it gave me something to think about. The reason I had never pondered it at great length was because I thought there was no way of knowing. I thought more about her question, and the Jaredites coming through the Orient, and wondered if that might be a key to learn what they looked like. With this in mind, I started my search and began coming across some things that might give us some insight.

First, of course, is to consider the scriptures. If they were not all destroyed, as previously mentioned, they would have been among the Mulekites and Nephites.

Mosiah 25:13 And now all the people of Zarahemla were numbered with the Nephites, and this because the kingdom had been conferred upon none but those who were descendants of Nephi.

So, what does the Jaredites among the Nephites have to do with what they might have looked like? As previously mentioned, Moroni, or

Moroni taking from the words of Ether, tells us they were fair. Speaking of Coriantumr, the scripture says:

Ether 13:17 - But he repented not, neither his fair sons nor daughters; neither the fair sons and daughters of Cohor; neither the fair sons and daughters of Corihor; and in fine, there were none of the fair sons and daughters upon the face of the whole earth who repented of their sins.

Nibley tells us the word "earth" and "land" are almost always used interchangeably [5]. So, this is probably referring to the land or country of the Jaredites.

We also have an interesting phrase with regard to their travel as previously mentioned.

Ether 2:5 And it came to pass that the Lord commanded them that they should go forth into the wilderness, yea, into that quarter where there never had man been.

This would mean that if they traveled through the Orient, they were the first ones there, at least after the flood. When the Lord takes a people through an unpopulated land, it seems he has at least some of them stay there to multiply and replenish the earth in that region. In addition, when a person considers that Jared kept asking his brother to request of the Lord that more people could join them, their group may have been large. This could mean that the Jaredite party was too large to all travel across the sea in the barges, especially with animals and supplies for nearly a year-long voyage.

So, if they were the first people in the region where they traveled, and that area was the Orient, could this give clues about what the Jaredites looked like? Could archeology add anything to the scriptures in this area? The answer is: only if we could find out the physical characteristics of the most ancient inhabitants of that land.

That might be difficult, because even in looking at old human remains that could be found, usually only skeletal fragments existed. A person wouldn't know hair color or complexion by them. Of course, today modern DNA testing might help, but if there was intermingling of nationalities, it could be convoluted. Another question is, could the people who are in the Orient now be their descendants?

These were tough questions to answer, especially since people have often been conquered, destroyed, or driven out of lands they first inhabited. That could have happened here. Tracing back to the first inhabitants would be difficult, and just looking at the people who live there now would be insufficient.

This all boils down to one thing. Is there any way to find out what the first people of the Orient looked like? And is there any way to prove whether or not the people living there now are descendants of those first people?

As I considered this, I thought of some things I had read from Thor Heyerdahl. He became increasingly interested in some hieroglyphics in Azerbaijan that he felt connected the Azerbaijanis to the Norwegian people [48]. He talked about how legends from Norway, where he was from, said that a man named Odin was the one who led people there.

Figure 37: Odin

A quick read of Mythopedia tells us that Odin was considered a god in the Norse legend and was also a wanderer [49]. You may not be as familiar with him as with Thor, the hammer-wielding deity. In some sources, Thor and Odin are brothers, and in others, Thor is Odin's son. But Heyerdahl believed Odin was real man, a tribal leader who led his people to the northern regions, and later his descendants created legends that raised him to the level of deity. Legends of Odin are not confined just to Norse folklore, either, but are also found extensively in Germanic folklore. He also appears frequently in stories from other lands. If he was a real person who led a people to those lands, the question was, where did they come from?

Azerbaijan is in the Caucasus region near the Black Sea and is on the edge of the orient. But Heyerdahl believed there could have been groups traveling west, north, and south who were connected to Odin. Much of his work, and the work of others, say they feel Odin's people traveled from eastern lands on their journeys to their new home or homes [48].

It is possible that instead of coming from the Caucasus region or even just east of the Caucasus region, as Heyerdahl proposed, the people of Odin came from even farther to the east and spread themselves from Azerbaijan to Norway and beyond. But why did they leave their homes and travel to new ones?

Hugh Nibley talks about the reasons people might uproot themselves and their families in large numbers [5]. The main reasons he gives are:

1. Need for food. This could be caused by famine, population growth, or man-made problems such as overgrazing.
2. Natural disasters. Nibley talks about tremendous storms, earthquakes, or other challenges that cause people to leave their current areas. He suggests that there are even stories of a great windstorm that caused the destruction of the tower of Babel and the dispersal of those there, including setting the Jaredite exodus into motion. This also might explain why they had to build ships tight like a dish to stand up to the ocean's waves.
3. War. People may be fleeing war or be driven away by invaders who themselves are being displaced.

I wondered whether Odin might have led people out of the Orient to European countries if he did. In particular, could the Jaredite descendants be the forebears of the Scandinavians, Azerbaijanis, Germans, and other European people? If so, we would know a lot about what they might have looked like by considering their descendants today. However, even though legend can be a basis for the beginning of an idea, a person needs more to create a foundation for a convincing argument toward a particular conclusion.

We do have Moroni's account that the Jaredites were a "fair" people. But does that necessarily mean they were possibly the red-headed or blond people we now find in many of these countries? "Fair" is a personal interpretation, as previously mentioned, and could mean many things, including simply that they were beautiful. I felt I needed more information to really consider this as a feasible idea.

I started searching for material about archeology work in the Orient, but nothing I found related to people as far back as the time of the Jaredites. And further, the features of being "fair" relate primarily to skin or hair. These characteristics decay and cannot be distinguished on skeletal remains.

Then I came across some fascinating material. In the first part of the nineteen hundreds, explorers came across mummies in the Tarim Basin, a desert where conditions existed that left the skin and hair of the mummies intact. Aurel Stein was one of the best-known of these explorers [50].

What made the mummies unique was that they were often tall. These mummies in the Tarim Basin are among the oldest ones found in China. In addition, the arid, hot sandy conditions there lent themselves to preserving skin and hair, probably better than the best embalming techniques of any ancient culture. The ancient Egyptians knew that salt was an anti-bacterial agent and a preservative and used it heavily in embalming. The Tarim Basin has a naturally salty atmosphere, providing natural protection to the bodies.

What was striking about these well-preserved mummies was the color of their hair and skin. Some of them were blond or red-haired and fair-skinned. Some could probably be 6 feet tall or more [51], though some question that. The dating techniques put them as far back as 2000 BC. The Jaredites and the Tower of Babel were probably around 2400-2000 BC. Interestingly, the mummies at the lower levels, the oldest ones, are more of the type that were tall and red-headed or blond. Those on the higher levels, which would be from more recent centuries, look more like those who live in the Orient now.

In his book *The Early Empires of Central Asia*, published in 1939, McGovern claims from his research and in-depth knowledge of the Chinese people, that the Indo-Europeans came from central Asia [52]. Breasted, in his book *Ancient Times a History of the Early World*, published in 1916, claims that these Indo-Europeans were the Scythians "the great white race" [53]. The idea of the Asian lands having white people seems to be common in many early books, but as we move further into the 1900s, the idea of a white race preceding those who are there now faded.

Though books were making this claim, and though the discovery of the mummies was groundbreaking, at that time men like Stein had other things to draw their interest. Little attention was paid to the mummies until 1988 when Victor Mair, a sinologist (expert in Chinese language, literature, history, society, etc.), came across some of these mummies in a backroom of a museum. His fascination and writing about them opened doors, and soon, others became interested in these mummies.

Besides their light skin and the color of their hair being unusual for that area, some of their clothes were as well. Their clothing included kilts with weaves that are like those in Scotland. Some experts believe these clothes hint at a migration from the steppe lands toward Europe, and not the other way around. According to textile experts, such as Elizabeth Barber, these textiles show that the people of Europe and their ideas, such as woven fabric, came from a people moving from the East to the west [50]. The weave is almost identical to cloth from Scotland, and weaves are unique and strong indicators of culture.

Early DNA testing showed that these mummies have strong tendencies toward matches with Europeans [50]. In other words, these more ancient mummies tend to match the appearance of the people of Europe, such as Germany or Scandinavia, and their DNA shows they are more closely related to the people of Europe than they are to the people who more commonly inhabit the Orient now.

An even more recent analysis is also interesting. An article in a recent National Geographic magazine says that the people of Europe had three different migrations. Through DNA testing of ancient human remains from other areas and comparing them to Europe, scientists have discovered the three waves of European immigrants [54]. The first one has been talked about much in science. That is that the first people came out of Africa and spread to all lands. According to scientists, the second wave was from the Middle East. But the final wave, which DNA science puts as in the last 5,000 years, has DNA matching remains in Russia and farther to the east. This would put this migration around 3,000 BC. (Note that the Jaredites would have been around 2,500 BC.) This would also mean that the remains of many in the Orient whose hair and skin were not preserved likely had hair and skin colors matching the Europeans, based on DNA.

What is my final analysis of what the Jaredites looked like? If the Jaredites were the first people in the Orient, then it appears likely that if there were those of the Jaredites left behind to populate those lands, they likely were the predecessors of modern-day northern Europeans. That would mean the Jaredite descendants later became those people. From all this analysis, I feel that the Jaredites looked much like the Vikings, Scandinavians, Germans, and those of the British Isles. We, of course, do not know this for sure, but there is a lot of evidence that it could be the case.

This leads us to one last question. If they were first in the Orient, what forced the Jaredite descendants out of those lands? This takes us back to what Hugh Nibley says about the events causing great migrations [5]. He says that great cataclysmic events often caused people to move. They would take over the lands of those who stood in their way, and those people would then move, encroaching on the lands of others, eventually forcing them out or destroying them.

Such an event, and this type of chain reaction, could have forced the Mongols or the Huns, as we know much of the people of the Orient today, to move east. This could have forced those in the east to move north and west. This would match what the legend of Odin says, as well as what much of the DNA and archeology suggest as possibilities.

With books from the early 1900s taken from Chinese manuscripts such as by Breasted [53] and McGovern [52] stating that central Asia was where the great white race came from, why were the mummies such a surprise? During the 1900s, especially with World War II occupations of China, and other strife, there were many upheavals in the societies of Asia. This seemed to have created nationalism and pride in the past and their culture. According to Mallory, Mair [50], and others, promoting ideas that other civilizations had preceded the current inhabitants of these lands had to be treated carefully not to antagonize the people and the government leaders.

But it is quite clear from the manuscripts, modern day findings, and technology that the possibility of a white race preceding the current inhabitants in the Orient is not only plausible but likely.

I, by no means, feel this analysis is exhaustive on the subject. But I do hope that it gives the reader a basis for doing their own research and coming to their own conclusions.

(Note: An interesting documentary, which includes information on the Tarim mummies, is the Lost Treasures of the Silk Road by Xterra. This can be found in the documentary section of Amazon videos.)

11
Sealed Portion of the Book of Mormon
+

I may be the only one, but I didn't catch the whole meaning of what was happening in the end of Ether 3, along with Ether 4 and 5, in my early readings of The Book of Mormon. Ever since I was young and had read about The Book of Mormon, I wondered what was on the sealed portion of the gold plates. But somehow, I never quite put it together that this was what Moroni was talking about in these sections, even though I had read the Book of Mormon many times. It seems obvious to me now, but I still think that insights into the Jaredites would not be complete without mentioning it in case others hadn't thought about it.

In Ether 3, the faith of the brother of Jared is so great that the Lord says he cannot keep him within the veil. He then not only shows himself to the brother of Jared, but He shows him all the inhabitants of the earth, past, present, and future. And he tells the brother of Jared to write down his vision, telling him to seal it up until the people were righteous enough and had faith like the brother of Jared.

Ether 3:25 And when the Lord had said these words, he showed unto the brother of Jared all the inhabitants of the earth which had been, and also all that would be; and he withheld them not from his sight, even unto the ends of the earth.

Moroni tells us that he had translated the brother of Jared's vision onto the gold plates and sealed them up as the Lord commanded.

Ether 4:4 Behold, I have written upon these plates the very things which the brother of Jared saw; and there never were greater things made manifest than those which were made manifest unto the brother of Jared. **Ether 4:5** Wherefore the Lord hath commanded me to write them; and I have written them. And he commanded me that I should seal them up; and he also hath commanded that I should seal up the interpretation thereof; wherefore I have sealed up the interpreters, according to the commandment of the Lord.

This means the sealed portion of The Book of Mormon contains

the complete vision of the brother of Jared. Nephi sees the translation of
The Book of Mormon in the last days. He knows the content of the
sealed portion and that it won't come forth until the Lord decides.

2 Nephi 27:10 But the words which are sealed he shall not deliver,
neither shall he deliver the book. For the book shall be sealed by the
power of God, and the revelation which was sealed shall be kept in the
book until the own due time of the Lord, that they may come forth; for
behold, they reveal all things from the foundation of the world unto the
end thereof.

There are people who have seen them. Moroni tells us that the
words were unsealed after Christ came to the Americas and the people
became righteous.

Ether 4:1 And the Lord commanded the brother of Jared to go down out
of the mount from the presence of the Lord, and write the things which
he had seen; and they were forbidden to come unto the children of men
until after that he should be lifted up upon the cross; and for this cause
did king Mosiah keep them, that they should not come unto the world
until after Christ should show himself unto his people.
Ether 4:2 And after Christ truly had showed himself unto his people he
commanded that they should be made manifest.

But after the people became wicked again, these writings were
once more sealed. Right in the book of Ether, Joseph Smith is
commanded not to touch the sealed portion.

Ether 5:1 And now I, Moroni, have written the words which were
commanded me, according to my memory; and I have told you the things
which I have sealed up; therefore touch them not in order that ye may
translate; for that thing is forbidden you, except by and by it shall be
wisdom in God.

It is to come forth in the last days when people became righteous
again.

2 Nephi 27:11 And the day cometh that the words of the book which
were sealed shall be read upon the house tops; and they shall be read by
the power of Christ; and all things shall be revealed unto the children of

men which ever have been among the children of men, and which ever will be even unto the end of the earth.

We are also told that the writings of John shall also come forth at that time.

Ether 4:16 And then shall my revelations which I have caused to be written by my servant John be unfolded in the eyes of all the people. Remember, when ye see these things, ye shall know that the time is at hand that they shall be made manifest in very deed.

This would be the revelation that Nephi talks about that he was not supposed to reveal since that assignment was given to John.

1 Nephi 14:25 But the things which thou shalt see hereafter thou shalt not write; for the Lord God hath ordained the apostle of the Lamb of God that he should write them.
1 Nephi 14:26 And also others who have been, to them hath he shown all things, and they have written them; and they are sealed up to come forth in their purity, according to the truth which is in the Lamb, in the own due time of the Lord, unto the house of Israel.
1 Nephi 14:27 And I, Nephi, heard and bear record, that the name of the apostle of the Lamb was John, according to the word of the angel.

So we know, these two records, the record of John the apostle and the writings of the Brother of Jared, will come out together, and there may be others. But how big is this record? The following quotes appear in an article by Kirk B. Henrichsen [55].

"What there was sealed appeared as solid to my view as wood. ... About the half of the book was sealed." David Whitmer

"A large portion of the leaves were so securely bound together that it was impossible to separate them." David Whitmer

"About two-thirds were sealed up, and Joseph was commanded not to break the seal; that part of the record was hid up. The plates which were sealed contained an account of those things shewn unto the brother of Jared." Orson Pratt

So, half to two-thirds of the book was sealed. Considering that the lost 116 pages, the book of Lehi, was part of the unsealed portion, it is valid to assume that the sealed portion could be quite a bit bigger than the portion we have. That will be an exciting day when we receive those writings. But I have heard more than one of the brethren ask why the Lord would bring forth that much new scripture if we aren't reading what we have. That is a good question, and I know I can do better.

12
Back to the Beginning – More than Academic Study
✦

I hope you have enjoyed this journey through the book of Ether. There is one final idea I would like to share as we end this time together.

A core concept of this book has been that the spiritual journey a person can gain from academic study of scriptures is more important than academic understanding. Both are important, and when they are coupled together, greater insight can be gained than either can give individually.

There is also a third, underlying thought that I would like to share. True testimony of God's work and the scriptures can only come from God and will never be fully realized outside of the ways He has prescribed. Although an academic study of the scriptures is good and should be done, it is insufficient. As a conclusion to this book, I would like to share an experience I had that brought this to the forefront of my understanding.

I had worked to complete a Bachelor of Arts education degree in mathematics at BYU. I planned to become a high school math teacher and a coach. But as I approached my final semester, the student teaching portion of the degree, I felt a strong prompting that I should change to a Bachelor of Science degree and do more math. With that prompting, I also felt that I should go on for an advanced degree, and that it had something to do with future work at Ricks College (now BYU – Idaho).

I had dreamed of teaching at Ricks but thought the chances were small. The math department faculty were all fairly young, and they hadn't had a new hire in years. Following the prompting, I looked at programs in the math fields I liked. I eventually settled on numerical analysis. This is a field that combines both math and computer science. I found two places that fit well. One was at Texas A and M University, and one was at Utah State University.

The move to Utah State, as well as the tuition, would be cheaper, and following direction I felt, I chose to go there. I was accepted and sent all the material for application to be a teaching assistant. I sent everything certified mail to guarantee its arrival, and I received an initial response that I would receive the position.

I moved my family to Logan, Utah, and went to meet with the professor who was over the teaching assistants. I will call him Dr.

Caldon. When I walked into his office, he said something that shocked me. He said, "I don't like you."

I had never met the man in my life, and I asked him why he didn't like me. He said, "I just don't," and wouldn't say more. I had another surprise in that he said he did not receive the application for the teaching assistantship, though I had sent it certified, and someone had signed for it. He told me that if I still wanted to go to school there, I could teach two classes for a wage of three thousand dollars. This was an enormous blow, since the assistantship I expected was supposed to be six thousand dollars for teaching a single class.

Not knowing what else to do, I accepted. I found other work to fill the gap between my pay and the money my family needed. Though it was discouraging at the time, these jobs provided enriching experiences, which are stories for another time. But needless to say, my beginning of work in graduate school was off to a rough start.

Later that fall, there was a party for graduate students, faculty, and their spouses at the home of a professor I will call Dr. Kevin Bozick. I had met Dr. Bozick previously, but I knew little about him other than he was the youngest professor in the department and was also an atheist, as were many of the professors.

I decided it would be good to go to the party, and Donna, my wife, agreed to go with me. When we arrived at Dr. Bozick's home, a sign on the door told the guests to go through the side gate into the backyard. Donna and I did this, and the first person we ran into was Dr. Caldon. He was drunk, and the minute he saw us, he started to mock me.

"Oh, look," he called out to the few other people who were already there. "The party is now ruined. We have a Mormon goody-two-shoes here, and we won't be able to have any fun."

He continued in this manner, and Donna, whispered, "What is he doing?"

I told her to ignore him and suggested we go to the opposite corner of the yard. We moved as far away from Dr. Caldon as possible, but that just made things worse. Instead of just speaking the things he had been saying, he yelled them out across the yard to make sure we could hear him. The only other people in the yard besides Dr. Caldon and us were a few graduate students, and what Dr. Caldon was saying seemed to annoy them, too. They all joined us to get as far away from Dr. Caldon as they could.

After Dr. Caldon had mocked me for a while, a group of the other professors entered the yard, led by Dr. Bozick. They were toting big coolers. When Dr. Bozick, being the host of the party, heard the things Dr. Caldon was yelling, he immediately told him to shut up. Then Dr. Bozick made an announcement.

"The coolers are all full of beer provided by the department. Everyone, feel free to drink all you want."

At that point, all the students with us moved across the yard, leaving Donna and me alone. Dr. Bozick came over. "Hey," he said, "why don't you come join us? Can I get you a beer?"

"Thanks," I replied, "but we don't drink alcohol."

Dr. Bozick looked surprised. "Why not?"

"We're members of the Church of Jesus Christ of Latter-Day Saints," I answered.

"You're what?" he asked.

"People call us Mormons," I replied.

Dr. Bozick stared at us for a moment in stunned silence. Then he asked, "Is not drinking alcohol a Mormon belief, or is it just yours?"

"It's a Mormon belief," I replied.

"I am so sorry," he said. "I didn't know. I have taught here for three years, and all the Mormon kids who have come through drank with us. None of them said anything to the contrary." He thought for a moment and then said, "Hey, I know. I've got Pepsi and Coke in the fridge. How about one of those?"

"That's kind," I said. "But many of us don't drink caffeine, including my wife and me. But water would be fine."

He shook his head. "Not at my party. It is not right for us not to have something for you. What would you normally drink at a party?"

"Well," I said, "I normally drink a root beer, and Donna drinks a Sprite."

He said nothing more but just nodded and turned around. He left the party, leaving his wife as hostess, and disappeared for about twenty minutes. When he returned, he had a root beer and a sprite for us. He must have gone clear to the store, which impressed me. Knowing Dr. Bozick was an atheist and respected my beliefs was a powerful contrast to Dr. Caldon's mocking of me.

The party was a luau, and there was wonderful roast pork, pineapple, and other great food. For the most part, we had a great time.

At every party I went to after that, and I'm sure even at the ones I didn't attend, when the coolers were opened, there was always a root beer and a sprite with tape around them and my name on the tape. Donna would never go with me again, but I would tell her about it. I knew that it was Dr. Bozick that had made sure it was there, and I appreciated it.

The following year, I took a class from Dr. Bozick. I found him to be an interesting teacher. One of the first things I learned was that he was close to my age, around twenty-four years old. He had graduated from high school at around sixteen and went straight through college to get a doctorate in mathematics.

Math was Dr. Bozick's world. He would really get absorbed in it. On our first day of class, he informed us that it was impossible to teach anything of value in a math class in one hour. So, instead of having class from three to four on Monday, Wednesday, and Friday, we were moving it to Tuesdays and Thursdays from four to five-thirty. It was still three hours per week, and it fit everyone's schedule, so we made the change.

Another idiosyncrasy that Dr. Bozick had was that he would walk into class with a couple of boxes of chalk. He would grab a new piece and start to lecture. As he did, he would get so excited about the math he was teaching that he would unknowingly break the chalk to pieces snapping it in his hand. Then when he turned to the board to write, he would see the small pieces of chalk in his hand. He hated small pieces of chalk, so he would toss them into the chalk tray and grab a new one. But when he turned to talk to us again, he would snap that new one into pieces as well. He went through both boxes of chalk each class period.

Another problem he had was that he would get so wound up in the math that nothing else mattered. He never wore a watch, and the room we were in was the department meeting room, which didn't have a clock. He would go and go and go, hardly taking a breath. And when he finally did, he would ask, "Anybody got the time?"

By the time he finally asked, it was almost always around nine o'clock. He would gasp when we told him since we were supposed to end at five-thirty, and he would say, "Is it really?" We would all smile at this question, because the room we were in stuck out from the building and had windows on three sides. The sun had been down for hours.

Once we assured him that it was indeed nine o'clock, he would say, "Let me finish this one last thing, and then we'll quit." After that, he would go on for another hour.

The first couple of times that I got home at eleven o'clock, I told Donna I was sure it was an anomaly, and he would start teaching for the expected hour and a half. But that never happened. I taught a class at seven every morning, so I was always exhausted when Dr. Bozick let us out.

The class was a full-year class going from early September to early June. We were less than a month from graduation when something interesting happened. As we ended at the typical ten o'clock time that night, Dr. Bozick called me by name and asked me to stay. He said he wanted to talk to me.

I stayed, and he said nothing until everyone else left. Once everyone else was gone, he asked, "Will you please close the door?"

I closed the door and returned to my seat, feeling somewhat anxious about what he wanted to talk about. He sat on the edge of his desk at the front of the room, folded his arms, and the first words out of his mouth caught me by surprise.

"There are some things about you that really bug me."

I was dumbfounded by this and couldn't speak for a moment. I had thought that we had become friends over that year. But finally, after I regained my composure, I said, "What is it, Kevin?"

This was another unusual thing about Dr. Bozick. He insisted we all call him by his first name. He didn't care for titles, and since he was about the same age as we were, calling him Dr. Bozick made him feel old.

"Do you remember that party at my house the fall you first came here about two years ago?" he asked.

"Yes, I remember," I replied.

"Well," he said, "when you wouldn't drink alcohol with us because of your religious beliefs, I said to my wife, 'I hope that man never takes my class because he couldn't be very logical.' I didn't want anyone lacking in logic in my class because I thought it would take away from the other students. And this last fall, when you signed up for my class, I felt the dread of having you in it.

"But it hasn't been like I thought it would be at all. In fact, you are my best student and the most logical person I have ever met. Your mathematical logic is impeccable. I would even say, you are more logical than I am at times."

"So, what's the problem?" I asked.

"The problem is that those two things, logic and a belief in God, don't go together in my mind. I keep thinking that your belief in God will gradually fade away, but it seems to get stronger, as does your logic. I have watched you, analyzed you, and evaluated every assignment you have written and done, but I don't feel any closer to understanding you than I did when the school year started. I have reread each paper and assignment you turned in at least three times, thinking someday I will find something that will be the key to understanding how this contradiction I see in you can be possible. But I find nothing. If I don't figure this out before the school year ends, I'm going to go crazy thinking about it. I want to know what makes you tick."

It was late at night, I was exhausted, and I had never had a question like this before, even in the mission field. I didn't know what to say.

"Kevin, can I have some time to think about this?"

He nodded. "Okay. But if you don't help me understand, I'll fail you."

I agreed that I would do my best to answer his concerns. We set a meeting for one week later after class to talk again.

I went home that night and told Donna what had happened. "He's not trying to be mean," I said. "He's really sincere and wants to know. And you remember how kind he was to us at that party a couple of years ago."

I told Donna I didn't know how to answer Dr. Bozick and asked her what she thought I should do.

"Pray about it," she said.

I sarcastically thought, "Thanks for the help," but that's exactly what I needed to do.

That was a Thursday, and when the weekend came, I fasted and prayed to know what to do and say. That Sunday morning, I had a strange experience. Donna and I went to church, and she went ahead of me to take our oldest daughter to the nursery. I followed a minute after carrying our baby and the diaper bag.

But everything was strange. Usually, the foyer would be full of people visiting, but this time there was only one old gentleman there. I looked at my watch, and it was the normal time when the foyer would be bustling with people conversing. My first thought was that it must be stake conference week, and we were missing it.

As I stood there, feeling somewhat confused, the old man walked over to me.

"You're a mathematician, aren't you, young man?" he asked.

"Yes," I replied.

"I have something you need to read," he said, and handed me a paper.

He turned and walked away. I looked at the paper, and it was a talk by Dallin Oaks called, "In the Lord's Way," which I think he had given at BYU when he had been president there. When Donna came

142

back, I asked her who the old man was who was in the foyer. I figured she would know because she knew almost everyone within about a month of moving into a new ward. On the other hand, it took me about a year to find out the bishop's name.

"What old man?" Donna asked.

"The one that was in the foyer," I replied.

"I don't remember," she said.

"Don't you think that is was strange that the foyer was so empty?"

She thought a moment and then said, "Come to think of it, that was strange."

I'm still not sure who the man was. I think I might have seen him around the ward after that. But the critical thing was the talk he had given me, and I'm grateful he shared it.

In the talk, Elder Oaks said that there are things we learn in the ways of science, and there are things we learn in the ways of God. To learn the things of science, we propose a hypothesis and test it, then adjust as necessary and test again. But to learn the things of God, we must do it in God's way. That includes reading scriptures, prayer, and fasting.

I thought a lot about this and realized it was the answer I needed. I reread the talk multiple times and considered how I would explain it to Dr. Bozick. When the next Thursday came, Dr. Bozick let class out early, like nine o'clock, then invited me to go to his office. When we got there, he sat in his chair, leaned back, put his feet on his desk, folded his arms, and said, "Well?"

"Kevin," I said, "this is one of the hardest things I have ever done, and the reason is because you don't have any of the prerequisites to understand what I am going to tell you. It would be like us going out on the street and dragging the first person we see into our math class, telling them they just had to see the math. They wouldn't understand any of it and would think we were crazy. But I will try my best."

I then told him about the talk by Elder Oaks, talking about how the things of science had to be learned in the ways of science, and the things of God had to be learned in God's way. I explained, as Elder Oaks did, that a person needed to study science in the ways of science and to seek truths about God in God's way. I told him that neither was sufficient for the other.

When I finished, I said, "Basically, Kevin, it boils down to this. There are things I know with my mind because I have proven them with logic. But there are things I know at least as well with my heart because I have proven them in God's way, and I do know them."

Dr. Bozick leaned back in his chair and stared at the ceiling for some time, deep in thought. Finally, he started to nod. Eventually, he looked at me and said, "That makes a lot of sense. My father was an atheist, as was my grandfather. I have never sought to find God because I have always been taught that he doesn't exist. But it makes sense that if he does exist, a person would have to find him in a different way."

The experience with Dr. Bozick was many years ago. He is no longer at Utah State University, but I see him now and then at math conferences. Sometimes he will still ask me religious questions, and we have wonderful discussions.

As important as the question about God and logic was to him, I feel it was more important to me. I had to really think about how and why my testimony of God was what it was and is what it is now.

I also realized that intellect alone can never be a basis for a real testimony. Over the years, I have seen some of my friends try to prove the gospel by pure intellect. They have never succeeded. In fact, most of the time, they have fallen away from the church.

I realize that a person never will be able to prove the gospel by intellect alone. One of the main purposes we have for being here on earth is to live by faith. If we try to prove the gospel academically, we will always hit holes. And God, as our loving father, will not fill in all the holes because he knows how important it is for us to live by faith and trust him.

Over the years, as I have thought of this experience, I have also considered some other things that have concerned me. One was that Dr. Bozick had taught at Utah State University for three years and all the students who were members of The Church of Jesus Christ of Latter-Day Saints always drank alcohol with everyone else. There were many students who were members of the church at the university. And in all the time I was there, for all the parties I attended, I always stood alone. It became easier with Dr. Bozick making sure there was a non-alcoholic drink for me. And over time, everyone learned I would not take part in drinking alcohol with them and respected my decision.

One last problem dealt with Dr. Caldon, the professor that had mocked me at the party. One day I was sitting in my office studying when the department chairman's daughter, whom I will refer to as Sally, came into my office. Her office was right next to mine. She was also a graduate student working on her doctorate, and we had become good friends.

On this day she said, "I'm sick of studying; let's talk."

I put away my work, and we visited. She saw my scriptures sitting on my desk, and she started asking me religious questions. She said she had read the Book of Mormon, but there was a lot of it she didn't understand. She said what she understood opened her mind a lot. Then she mentioned that only three department faculty were members of the church. When she named them, one was Dr. Caldon.

"That can't be right," I said. "You've seen how he mocks me for my standards and what I believe."

She nodded. "I can't believe how openly he mocks you. I've never seen anything like it."

Then she said something I thought was very insightful. She said, "I think he mocks you because he sees in you what he should be, and since he's not, it makes him feel guilty, and he gets angry with you."

I have found that the most negative criticism I receive for what I believe and the standards I try to live comes from those who are, or have been, members of The Church of Jesus Christ of Latter-Day Saints who have turned away. My friends from other faiths are respectful of my beliefs as I always try to be of theirs.

In conclusion, I hope that as you search for understanding of the scriptures, that you always remember that intellectual knowledge is never sufficient. A true testimony can only come in God's way. So, although it is wonderful to gain academic understanding, always remember to also seek in God's way. I know he will always help you and answer you in his own way and his own time as he has done for me.

Bibliography

[1]   G. B. Hinckley, "God Has Not Given Us the Spirit of Fear," *Edited version of an address given 5 November 1983 to Latter-day Saint college students at the Salt Lake Institute of Religion*, 1983. https://www.lds.org/ensign/1984/10/god-hath-not-given-us-the-spirit-of-fear (accessed Nov. 23, 2022).

[2]   "Nuremberg Executions." https://en.wikipedia.org/wiki/Nuremberg_executions (accessed Nov. 23, 2022).

[3]   Sadam Hussein, "Sadam Hussein." https://en.wikipedia.org/wiki/Saddam_Hussein (accessed Nov. 23, 2022).

[4]   H. Murray, *A Brief History of Chess*. New York and Tokyo: Ishi Press, 2015.

[5]   H. Nibley, "The World of the Jaredites," in *The Collected Works of Hugh Nibley Volume 5*, Salt Lake City, Utah: Deseret Book Company, 1988, pp. 153–282.

[6]   N. Klimczak, "The Dramatic True Story Behind Disney's Mulan," 2020. https://www.ancient-origins.net/history-famous-people/ballad-hua-mulan-legendary-warrior-woman-who-brought-hope-china-005084 (accessed Nov. 23, 2022).

[7]   "Old Testament Chronology," in *Old Testament Study Manual*, Church of Jesus Christ of Latter-Day Saints, 2015. [Online]. Available: https://www.churchofjesuschrist.org/study/manual/old-testament-study-guide-for-home-study-seminary-students-2015/old-testament-chronology?lang=eng

[8]   "Book of Mormon Chronology," *Ensign*. https://www.churchofjesuschrist.org/bc/content/shared/content/images/gospel-library/magazine/ensignlp.nfo:o:334e.jpg

[9]   B. Hodge, "Was the Tower of Babel Dispersion a Real Event?," *Answers In Genesis*, 2010. https://answersingenesis.org/tower-of-babel/was-the-dispersion-at-babel-a-real-event/

[10]   T. Heyerdahl, *Kontiki*. Chicago: Rand McNally & Company, 1951.

[11]   "Who Begat Whom?" https://answersingenesis.org/bible-timeline/genealogy/who-begat-whom/ (accessed Dec. 23, 2022).

[12]   D. Ludlow, *Companion to Your Study of the Book of Mormon*. Deseret Book Company, 1977.

[13]   A. Habermehl, "Where in the World was the Tower of Babel?," *Answers In Genesis*, 2011. https://answersingenesis.org/tower-of-babel/where-in-the-world-is-the-tower-of-babel/

[14]   "Nimrod." https://en.wikipedia.org/wiki/Nimrod (accessed Nov. 24, 2022).

[15]   T. Heyerdahl, *The Ra Expeditions*. Garden City, New York: Doubleday & Company, 1971.

[16]   "Flotsam from 2011 Japan tsunami reaches Alaska," *CNN*, 2012. [Online]. Available: https://www.cnn.com/2012/05/22/us/alaska-tsunami-debris/index.html

[17]   D. H. Smith, "Divine Kingship in Ancient China," *Numen - International Review for the History of Religions*, vol. 4, no. 3, pp. 171–203, 1957, Accessed: Oct. 17, 2022. [Online]. Available: https://www.jstor.org/stable/3269343

[18]   Genghis Khan, "Genghis Khan." https://en.wikipedia.org/wiki/Genghis_Khan (accessed Oct. 23, 2022).

[19]   J. Weatherford, *Genghis Khan and the Making of the Modern World*. 2004.

[20]   "PAWN | definition in the Cambridge English Dictionary." https://dictionary.cambridge.org/us/dictionary/english/pawn (accessed Oct. 23, 2022).

[21]   "Two Crowns, Three Respects." https://simple.wikipedia.org/wiki/Er_Wang_San_Ke (accessed Nov. 03, 2022).

[22]   "Chinese Nobility." https://en.wikipedia.org/wiki/Chinese_nobility (accessed Nov. 03, 2022).

[23]   "Glass_Encyclopedia." https://www.encyclopedia.com/science/encyclopedias-almanacs-transcripts-and-maps/development-glassmaking-ancient-world (accessed Nov. 06, 2022).

[24]   "Steel." https://en.wikipedia.org/wiki/Steel (accessed Nov. 23, 2022).

[25]   "Silk." https://en.wikipedia.org/wiki/History_of_silk (accessed Nov. 06, 2022).

[26]   "Silk Worm Diseases."
         https://www.sciencedirect.com/topics/medicine-and-
         dentistry/silkworm-
         disease#:~:text=Major%20silkworm%20diseases%20are%20
         viral,BmDNV)%3B%20bacterial%20such%20as (accessed
         Nov. 06, 2022).

[27]   L. J. Johnston, "Yes world, there were horses in Native
         culture before the settlers came," *Indian Country Today*,
         2019. https://indiancountrytoday.com/news/yes-world-there-
         were-horses-in-native-culture-before-the-settlers-came
         (accessed Nov. 06, 2022).

[28]   M. Lenart, "Gone But Not Forgotten: Bring Back North
         American Elephants," *Science Daily*, 1999.
         https://www.sciencedaily.com/releases/1999/06/9906071543
         15.htm (accessed Nov. 06, 2022).

[29]   S. N. Kramer, *The Sumerians Their History, Culture, and
         Character*. The University of Chicago Press, 1963.

[30]   J. Grover, "Cureloms and Cumoms."
         http://scripturalmormonism.blogspot.com/2017/03/jerry-
         grover-on-cureloms-and-cummoms-in.html (accessed Nov.
         07, 2022).

[31]   Texas A&M University - Agricultural Communications,
         "Research Upsetting Some Notions About Honey Bees,"
         Science Daily, 2006.
         https://www.sciencedaily.com/releases/2006/12/0612112209
         27.htm (accessed Nov. 23, 2022).

[32]   "Insights into social insects from the genome of the honeybee
         Apis mellifera," *Nature*, 2006, doi: 10.1038.

[33]   "Succession to the Chinese Throne."
         https://en.wikipedia.org/wiki/Succession_to_the_Chinese_thr
         one (accessed Nov. 24, 2022).

[34]   "2008 Book of Mormon Seminary Study Guide," *Church of
         Jesus Christ of Latter-Day Saints*.
         http://media.ldscdn.org/pdf/scripture-and-lesson-
         support/book-of-mormon-seminary-student-study-
         guide/2008-01-000-book-of-mormon-seminary-student-
         study-guide-eng.pdf (accessed Jan. 21, 2023).

[35]   "Brother of Jared."
         https://en.wikipedia.org/wiki/Brother_of_Jared (accessed
         Dec. 20, 2022).

[36] "8 Facts About Ancient Egypt's Hieroglyphic Writing." https://www.history.com/news/hieroglyphics-facts-ancient-egypt (accessed Dec. 20, 2022).

[37] "Naming Taboo." [Online]. Available: https://en.wikipedia.org/wiki/Naming_taboo

[38] "Codex Name Generator and Backstories." https://codexnomina.com/sumerian-names (accessed Jan. 21, 2023).

[39] E. Chiera, "List of Personal Names from the Temple School of Nippura," *Babylonian Section*, vol. XI, no. 1, 1916, Accessed: Nov. 16, 2022. [Online]. Available: http://www.etana.org/sites/default/files/coretexts/14902.pdf

[40] "Sumerian Personal Names." https://www.theishtargate.com/personal-names.html (accessed Nov. 24, 2022).

[41] "Book of Jubilees." Accessed: Dec. 02, 2022. [Online]. Available: https://dl.icdst.org/pdfs/files1/79210c539e32f6c4f106fab6fff3f0c4.pdf

[42] "Native American First Names." https://www.familyeducation.com/baby-names/first-name/origin/native-american (accessed Nov. 24, 2022).

[43] "Sumerian King List." https://en.wikipedia.org/wiki/Sumerian_King_List (accessed Nov. 24, 2022).

[44] "Sumer." https://en.wikipedia.org/wiki/Sumer (accessed Jan. 22, 2023).

[45] "Sumerian Cuneiform and Names." https://en.wiktionary.org/wiki/%F0%92%88%A4 (accessed Jan. 22, 2023).

[46] "Word Reference." https://www.wordreference.com/definition/am (accessed Jan. 22, 2023).

[47] "Carchemish." https://en.wikipedia.org/wiki/Carchemish (accessed Nov. 24, 2022).

[48] T. Heyerdahl, "The Azerbaijan connection Challenging Euro-centric theories of migration," *Azerbaijan International*, vol. 3.1, no. Spring, pp. 60–61, 1995, [Online]. Available: http://azer.com/aiweb/categories/magazine/31_folder/31_articles/31_thorazerconn.html

[49] T. Apel, "Odin." Accessed: Apr. 22, 2023. [Online].
      Available: https://mythopedia.com/topics/odin

[50] J. P. Mallory and V. H. Mair, *The Tarim Mummies*. New
      York, New York: Thames & Hudson, 2000.

[51] C. Coonan, "A meeting of civilizations: The mystery of
      China's Celtic mummies."
      http://www.independent.co.uk/news/world/asia/a-meeting-of-
      civilisations-the-mystery-of-chinas-celtic-mummies-
      5330366.html (accessed Jul. 25, 2022).

[52] W. M. McGovern, *The Early Empires of Central Asia*. The
      University of North Carolina Press, 1939. [Online].
      Available:
      https://ia801601.us.archive.org/25/items/in.ernet.dli.2015.53
      5974/2015.535974.early-empires.pdf

[53] J. H. Breasted, *Ancient Times a History of the Early World*.
      Ginn and Company, 1916.

[54] A. Curry, "Genetic testing reveals that Europe is a melting
      pot, made of immigrants," *National Geographic*, 2019.
      [Online]. Available:
      https://www.nationalgeographic.com/culture/article/first-
      europeans-immigrants-genetic-testing-feature

[55] K. B. Henrichsen, "How Witnesses Described the Gold
      Plates," *Journal of Book of Mormon Studies*, vol. 10, no. 1,
      2001, Accessed: Nov. 24, 2022. [Online]. Available:
      https://scholarsarchive.byu.edu/cgi/viewcontent.cgi?article=1
      267&context=jbms

Appendix A

Table 3 - King Succession

When A and B kings are listed, the kingdom is split

Chp	King #	King	Captivity by and Relation	Killed by	Notes
7	1	Orihah			
7	2	Kib	Corihor (son)		
7	3	Corihor	Shule (brother)		
7	4	Kib (again)			
7	5	Shule	Noah (Nephew)		Noah almost killed Shule, but Shule's sons saved him and killed Noah
7	6A	Noah		Shule's sons	
7	6B	Cohor		Shule (Great Uncle)	Cohor killed by Shule in battle
7	6B	Nimrod			Gives up his part of kingdom back to Shule
7	6	Shule			Rules both parts again
8	7	Omer	Jared (son)		
8	8	Jared			Omer's sons killed rest of Jared's army
8	9	Omer (again)			Warned to flee, Akish tried to kill to get Jared's daughter for wife, would have been killed, comando kind of plot
9	10	Jared (again briefly)		Akish (no relationship given)	Commando type killing
9	11	Akish		From battle with sons	Battle killed all but 30 except those who had gone to join Omer
9	12	Omer (again)			
9	13	Emer			
9	14	Coriantumr			

9	15	Com		Heth (son)	Heth slew Com with his own sword
9	16	Heth		died in famine	
10	17	Shez			
10	18	Riplakish		by the people	descendants driven out
10	19	Morianton			Son of Riplikash, came back, fought people, and took over then eased taxes and found favor with people
10	20	Kim	By brother - not named		
10	21	Kim's Brother			Defeated by Levi, son of Kim born in captivity, but does not say if killed or captured.
10	22	Levi			
10	23	Corom			
10	24	Kish			
10	25	Lib			
10	26	Hearthom	not named – no genealogy given		
10	27	Com (next named ruler)			Com son of Coriantum son of Amnigaddah son of Aaron son of Heth, all who dwelt in captivity. Com got half of the people, then beat Amgid to get rest
11	28	Shiblom		Brother - not named	Says Shiblom slain and Seth (probably son) in captivity
11	29	Seth	not named – no genealogy given		
11	30	Ahah (Seth's son)			Aha obtains kingdom. It could be he is the one that put Seth in captivity but doesn't say.
11	31	Ethem			
11	32	Moron	Mighty man (no genealogy given)		

11	32B	Mighty man			Takes half of kingdom – later loses it, but does not say what happens to him
	32	Moron	Another mighty man (descendant of Brother of Jared)		Moron takes back full kingdom – then loses all and lives in captivity. Coriantor, son of Moron, dwelt in captivity and begat Ether
	33	Another mighty man			
12	34A	Coriantumr			In days of Ether
13	34B	Others - mentions Cohor, Corihor	No genealogy given		Much fighting from secret combinations
13	34A	Coriantumr	Shared (no genealogy given)		Nibley claims Shared and Coriantumr were "royal brothers", but I can't find anything in the scriptures to support that.
13	34B	Shared	Sons of Coriantumr	Coriantumr	Sons of Coriantumr beat Shared, but not sure about captivity. But they don't kill him, so he likely fled. But later Shared and Coriantumr battle and Coriantumr kills Shared
14	34B	Gilead, Brother of Shared		by his high priest	
14	34B	Lib		Coriantumr	Made king by secret combination
15	34B	Shiz (Brother of Lib)		Coriantumr	

Table 4 - Table of Figures

Table 5 - Picture Attribution and Links

#	Attribute to	Link
1	Александр Михальчук - Wikimedia Commons	https://upload.wikimedia.org/wikipedia/commons/1/1e/The_Tower_of_Babel_Alexander_Mikhalchyk.jpg
2	Pixabay	https://pixabay.com/illustrations/chess-strategy-game-king-board-1709600/
3	Anandajoti Bhikkhu from Sadao, Thailand - Wikimedia Commons	https://commons.wikimedia.org/wiki/File:019_Hua_Mulan_%2825596209557%29.jpg
4	Daris Howard	Personal Image
5	Lyaschuchenko, CC BY-SA 4.0 - Wikimedia Commons	https://commons.wikimedia.org/wiki/File:%D0%A0%D0%BE%D0%B7%D0%BA%D0%BE%D0%BF%D0%BA%D0%B8_%D0%95%D1%82%D0%B5%D0%BC%D0%B0%D0%BD%D0%B0%D0%BD%D0%BA%D0%B8.jpg
6	Google Maps	Etemenanki Google Map Search
7	Entemenanki Middle East Map	Google Maps edited
8	Robert Koldewey - Wikipedia	https://commons.wikimedia.org/wiki/File:Etemenanki_Babylon_(3).png
9	Robert Koldewey - Wikipedia	https://upload.wikimedia.org/wikipedia/commons/9/9c/Etemenanki_Babylon_%283%29.png
10	U.S. CIA - Public domain - Wikimedia Commons	https://upload.wikimedia.org/wikipedia/commons/7/70/Middle_East._LOC_2001625293.jpg
11	Pixabay	https://pixabay.com/illustrations/map-map-of-the-world-relief-map-221210/
12	Pixabay - Edited	https://pixabay.com/illustrations/map-map-of-the-world-relief-map-221210/
13	Daderot - Wikimedia Commons	https://commons.wikimedia.org/wiki/File:Kon-Tiki_raft_-_IMG_9232.jpg
14	GAD - Wikimedia Commons	https://upload.wikimedia.org/wikipedia/commons/b/b4/Ra_2.jpg
15	Daris Howard	Personal Image

16	Daris Howard	Personal Image
17	Daris Howard	Personal Image
18	Daris Howard	Personal Image
19	KVDP - Wikimedia Commons	https://commons.wikimedia.org/wiki/File:Map_prevailing_winds_on_earth.png
20	Chi King - Wikimedia Commons	https://commons.wikimedia.org/wiki/File:Huangshan_pic_4.jpg
21	Pixabay	https://pixabay.com/photos/yellow-mountains-mountain-ranges-532857/
22	Kanguole - Wikimedia Commons	https://commons.wikimedia.org/wiki/File:China_Northern_Plain_relief_location_map.png
23	NOAA - Wikimedia Commons	https://upload.wikimedia.org/wikipedia/commons/8/8a/Oceanic_gyres.png
24	Bill Taroli - Wikimedia Commons	https://upload.wikimedia.org/wikipedia/commons/7/7d/Mural_of_siege_warfare%2C_Genghis_Khan_Exhibit%2C_Tech_Museum_San_Jose%2C_2010.jpg
25	KoizumiBS - Wikimedia Commons	https://en.wikipedia.org/wiki/Jamukha#/media/File:Jamukha_&_Toghrul.jpg
26	Pixabay	https://pixabay.com/photos/cocoon-silkworm-silk-cut-open-196533/
27	Pixabay Multiple Images	Images: 6464231, 5836459, 1730075, 4275741
28	Pixabay	https://pixabay.com/photos/peru-the-colca-valley-inca-andes-2363496/
29	Pixabay	https://pixabay.com/photos/nopal-tunas-cactus-cacti-thorns-6035102/
30	Jolle at Catalan - Wikimedia Commons	https://commons.wikimedia.org/wiki/File:Orientmitja2300aC.png
31	Daris Howard	Personal Image
32	Pixabay	https://pixabay.com/photos/bees-hive-insects-macro-honey-bees-326337/
33	Hosni bin Park - Wikimedia Commons	https://upload.wikimedia.org/wikipedia/commons/b/bd/Egyptian_hieroglyphics.jpg
34	Mary Harrsch - Wikimedia	https://commons.wikimedia.org/wiki/File:Cuneiform_tablet_in_the_name_of_Shar-Kali-Sharri.jpg

35	Wikimedia Commons - Transcript: Stephen Herbert Langdon	https://commons.wikimedia.org/wiki/File:Weld-Blundell_Prism_with_transcription_by_Stephen_Herbert_Langdon_(1876-1937).jpg
36	Daris Howard	Personal Image
37	Georg von Rosen - Wikimedia	https://commons.wikimedia.org/wiki/File:Georg_von_Rosen_-_Oden_som_vandringsman,_1886_(Odin,_the_Wanderer).jpg

www.ingramcontent.com/pod-product-compliance
Lightning Source LLC
Chambersburg PA
CBHW070543090426

42735CB00013B/3056